ISBN 978-1-332-33734-7
PIBN 10315878

1 MONTH OF
FREE
READING

at

www.ForgottenBooks.com

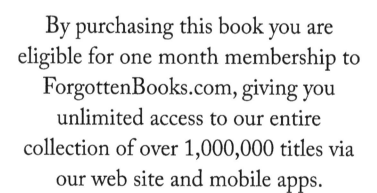

By purchasing this book you are eligible for one month membership to ForgottenBooks.com, giving you unlimited access to our entire collection of over 1,000,000 titles via our web site and mobile apps.

To claim your free month visit:

www.forgottenbooks.com/free315878

English
Français
Deutsche
Italiano
Español
Português

www.forgottenbooks.com

Mythology Photography **Fiction**
Fishing Christianity **Art** Cooking
Essays Buddhism Freemasonry
Medicine **Biology** Music **Ancient**
Egypt Evolution Carpentry Physics
Dance Geology **Mathematics** Fitness
Shakespeare **Folklore** Yoga Marketing
Confidence Immortality Biographies
Poetry **Psychology** Witchcraft
Electronics Chemistry History **Law**
Accounting **Philosophy** Anthropology
Alchemy Drama Quantum Mechanics
Atheism Sexual Health **Ancient History**
Entrepreneurship Languages Sport
Paleontology Needlework Islam
Metaphysics Investment Archaeology
Parenting Statistics Criminology
Motivational

THE

Physiology of Alimentation

BY

DR. MARTIN H. FISCHER

Professor of Pathology in the Oakland College of Medicine

FIRST EDITION

FIRST THOUSAND

JOHN WILEY & SONS

London: CHAPMAN & HALL, Limited

1907

ROBERT DRUMMOND, PRINTER, NEW YORK.

TABLE OF CONTENTS.

CHAPTER I.

CHAPTER II.

CHAPTER III.

PHYSIOLOGY OF ALIMENTATION.

CHAPTER I.

THE MECHANICAL PHENOMENA OF ALIMENTATION.

1. The General Functions of the Alimentary Tract.—
Under the functions of the alimentary tract are included all the
functions of the hollow tube which begins with the mouth
and ends with the anus, together with certain of those of
the glands which pour their secretions into this tube. The
changes which the food undergoes in its passage through
this tube are in part purely *mechanical*—such for example
as are the consequence of mastication or the movements
of the stomach—in larger part, however, *chemical*. As an
example of the latter may be mentioned the conversion in
the stomach of albuminous bodies such as egg white into
the chemically less complex peptones. Yet these chemical
changes are frequently associated with physical or physico-
chemical ones which from a physiological standpoint may
at times be quite as important as the chemical changes
themselves.

We can classify the various substances which serve as
food under the general headings of proteins, carbohydrates,
fats, and inorganic substances. Under the first heading
fall, for example, the lean meats, the white of egg, etc.,
while the chief representatives of the carbohydrates are
the starches and sugars which we consume. The fats are

in part of animal, in part of vegetable origin. Among the vegetable fats are the cottonseed and olive oils, while familiar examples of animal fats are found in butter, cream, and the fats of fat meat. Water makes up the bulk of the inorganic material which we consume. Among the other inorganic constituents of our food are the all-important salts, which come to us in part as natural components of our diet, in part as condiments, or as constituents of the so-called mineral waters.

Representatives of all these classes are found in the ordinary mixed diet, and no diet is sufficient for man for any length of time unless it contains representatives of all these classes. This mixed diet enters the alimentary tract, sometimes finely divided, sometimes in a state of coarse division. Some of the constituents of the food may be in suspension or in solution in water. At times the food is cooked, at other times uncooked, and in this state is started on its way through the alimentary tract. The alimentary tract takes up or absorbs from this heterogeneous mass which enters the mouth certain constituents of the food either in the condition in which they are consumed by the individual or after they have been acted upon by the secretions of the alimentary tract. Besides altering the physical and chemical constitution of the food the alimentary tract therefore acts as an *absorptive system*. Interestingly enough, however, it also acts as an *excretory system*. All these various functions, not only the elaboration of the food by physical and chemical means, but also those of absorption and excretion, are included under the general caption, *alimentation*.

We shall now take up these functions separately, directing our attention first to the *mechanical* phenomena which are associated with the journey of the food through the alimentary tract.

2. Mastication.—The articulation of the inferior maxillary bone with the skull allows of a variety of movements all of which are under the control of the will. The lower maxillary bone may be dropped and raised, may be thrown forward and drawn backward, and may be moved from side to side. Ordinary mastication in the human being is a combination of all these movements. The lower jaw is lowered and raised in the ordinary biting movements, and moved from side to side when the food is being chewed. Combined with both of these may be more or less well-marked forward and backward movements of the jaw. The food is kept between the teeth and the act of mastication made more effective by the simultaneous action of the muscles of the tongue, cheeks, and lips. The cheeks, and more especially the tongue, aid also in gathering together the food in the mouth and forming it into a bolus preparatory to the act of swallowing.

The muscles concerned in the movements of the lower jaw are the following. The masseter, temporal, and internal pterygoid raise the jaw. The digastric is the chief depressor of the lower maxilla, aided at times by the mylo-hyoid and genio-hyoid muscles. The jaw is thrown forward by the simultaneous contraction of the external pterygoids. When these muscles move singly, side-to-side movements are produced. The jaw is retracted by contraction of the temporal muscle. The muscles of mastication receive their nerve supply from the inferior maxillary division of the fifth cranial nerve with the exception of the genio-hyoid, which is supplied by the hypoglossal nerve.

3. Deglutition.—It seems to be essential for the proper performance of the act of deglutition that the mass to be swallowed be moist. Dry material can either not be swallowed at all or at best with difficulty. While certain substances may therefore be swallowed immediately, it is necessary for others that they remain in the oral cavity until they have been thoroughly mixed with saliva, or, in people of improper dietary habits, until they have been moistened by admixture with a mouthful of water, tea, or other liquid.

The act of swallowing is usually divided into three parts corresponding to the anatomical regions through which the food has to pass, namely, the mouth, the pharynx, and the œsophagus. But this division, it will be seen, is a purely arbitrary one and therefore had best not be made. In its passage through the mouth and the upper portion of the pharynx the food may be kept under the control of the will. After the food has come into the grasp of the involuntary muscle fibres of the œsophagus its movement can no longer be controlled voluntarily. Under ordinary circumstances, however, with the exception of the voluntary formation of the bolus, the entire act is involuntary, and is essentially reflex in character.

The best experimental observations that we have on the act of deglutition are those of KRONECKER and MELTZER,[1] and those of CANNON and MOSER.[2] Although the experimental methods adopted by these investigators are radically different, their results agree in the main very well.

Preparatory to the act of deglutition the material to be swallowed is collected into a bolus through the combined movements of the cheeks, teeth, and tongue. The bolus rests for a moment on the dorsum of the tongue. According to KRONECKER and MELTZER the chief factor concerned in forcing food through the pharynx and œsophagus is the quick and powerful contraction of the mylo-hyoid muscles, aided by the simultaneous contraction of the hyoglossi muscles. The contraction of these sets of muscles puts the bolus of food as it rests on the dorsum of the tongue under high pressure and shoots it in the direction of the least resistance through the pharynx and œsophagus. By the contraction of these muscles the epiglottis is also closed over the tracheal

---

[1] KRONECKER and MELTZER: Archiv für Physiologie, 1880, p. 446; ibid., 1883, Suppl. Bd., p. 337, 351. MELTZER: Journal of Experimental Medicine, 1897, II, p. 457.

[2] CANNON and MOSER: American Journal of Physiology, 1898, I, p. 435.

opening. KRONECKER and MELTZER believe that the food passes in a spurt from the beginning of the pharynx clear through the œsophagus to the cardiac orifice of the stomach. The contraction of the constrictors of the pharynx and the peristaltic movements of the œsophagus, they believe, follow this act and serve to remove any fragments which may have adhered to the œsophagus. This description we shall see holds only for liquids, and not for more solid foods. They also believe that, in most individuals at least, the food does not immediately enter the stomach but is stopped at the lower end of the œsophagus by a contraction of the circular muscle fibres of the cardia, and is only slowly forced into the stomach by the aftercoming peristaltic wave. The experiments of CANNON and MOSER do not support this view.

CANNON and MOSER studied the act of deglutition in various animals, including man, by following the passage of liquids, solids, and semi-solids mixed with bismuth subnitrate from the mouth to the stomach by means of the x-rays. The bismuth subnitrate renders the swallowed mass opaque to the x-rays, and as this method necessitates neither anæsthetics, operative procedures, nor recording instruments it is freer from objection than some of the older means employed in the study of deglutition. According to these authors the movement of food through the œsophagus differs markedly not only in different animals but also in the same animal with food of different consistencies. In fowls, for example, the rate of movement through the œsophagus is always slow, and no matter what the consistency; it is carried from the pharynx into the stomach by peristaltic waves. A squirt-like movement when liquids are swallowed is impossible in these animals, as the parts forming the mouth are too rigid. In order to get the swallowed mass within the grasp of the œsophageal musculature the head is raised to aid the weak propulsive powers of the mouth as largely as possible by gravity..

In the cat also the food is moved through the œsophagus by peristalsis, but somewhat more rapidly than in the case of fowls. It requires nine to twelve seconds for a bolus of solid food to reach the stomach, and a somewhat shorter time for liquids to make the same journey. The reason for this difference lies in the fact that liquids move somewhat more rapidly in the upper portion of the œsophagus than do semi-solids. In the lower portion of the œsophagus the rate for both kinds of food is approximately the same. For all kinds of food the rate of movement in the upper half of the œsophagus is somewhat greater than in the lower half.

In the dog swallowing approximates the same act in the human being. The total time for the descent of a bolus of food in this animal is from four to five seconds. In the upper portion of the œsophagus the movement is always more rapid than in the lower, and when the swallowed mass is liquid this rapid movement continues deeper into the œsophagus than when it is solid or semi-solid. No distinct pause occurs when the bolus changes from its rapid rate to the slower one.

In man liquids are propelled deep into the œsophagus at a rate of several feet a second by the sudden and sharp contraction of the mylo-hyoid muscles. This confirms the observations of KRONECKER and MELTZER. According to CANNON and MOSER, however, solids and semi-solids are not swallowed in the same way. From studies on a seven-year-old girl who was given gelatine capsules filled with bismuth subnitrate, or bread-and-milk mush mixed with the same salt, they conclude that the movement of food of these consistencies through the œsophagus is always accomplished by peristalsis. Nor does the food in its passage through the œsophagus stop before it enters the stomach, which KRONECKER and MELTZER believed to be the case. Only the rate of progressive movement changes from a more rapid one in the upper œsophagus to a slower one lower down.

4. The Movements of the Stomach.—The movements of the various portions of the alimentary tract from the œsophagus to the rectum have been the object of research of many investigators for many years. The pages of this volume do not allow even an outline of the various views which have been held from time to time. Many of these we now know to be entirely false, others in part, often in large part, correct.

Within recent years a number of papers have appeared on the movements of the gastro-intestinal tract which have given us a clearer insight into this problem. Foremost among these newer researches stand the observations of CANNON in this country and ROUX and BATHAZARD in France. The observations of these men are free from many of the objections which may be lodged against the older studies of the subject.

Experiments in mammalian physiology must of necessity be so often carried on under anæsthetics or the disturbing influence of operative procedures that when these factors affect the physiological process which is being investigated results are obtained which if not wrong are at least confusing. It is to the disturbing influences of the methods used in the investigation of the movements of the gastro-intestinal tract by the older observers that some of their confusing results are to be attributed, and it is but natural that the introduction of experiments in which all operative interference, anæsthetics, etc., are shut out should yield more trustworthy results than our older ones; while they indicate to us at the same time what is right and what is wrong in our older conceptions.

CANNON and ROUX and BATHAZARD used the x-ray in their study of the movements of the alimentary tract. This does away with the necessity of surgical operations in order to obtain a view of the intestines, and at the same time prevents the exposure of the abdominal contents to the cold of the air, to evaporation, etc., all of them important factors in modifying the normal activity of the hollow viscera. In order to render the food visible within the alimentary tract, bismuth subnitrate was mixed with it. Since this substance is opaque to the x-rays, the food may readily be

followed in its passage through the alimentary tract by placing a fluoroscopic screen over the animal.

The only possible sources of error which might have crept
into these observations of CANNON and ROUX and BATHA-
ZARD are therefore those connected with tying the animal down
while making observations and the feeding of bismuth subnitrate with the food. That the first plays no rôle in carefully conducted experiments is indicated by the fact that
observations carried out on sleeping animals agree perfectly with those carried out on waking ones. So far as the
bismuth subnitrate is concerned the objection will be made
that it inhibits the intestinal movements, as it is used for
this purpose in the treatment of diarrhœas. But care must
be taken in applying what holds for an inflamed gastro-
intestinal tract to a healthy one, in which the action of the
bismuth subnitrate is at its worst but slight.

We shall follow first of all CANNON's [1] description of the
movements of the stomach in the cat.

The form of the active stomach a few minutes after a
meal of 15 grams of milk and bread mixed with 3 to 5 grams
of bismuth subnitrate is shown in Fig. 1. For convenience
in description the stomach may be divided into two parts.
The larger cardiac part lies to the animal's left of the line
through *wx*. The smaller pyloric part lies to the right of
this line and consists of two subdivisions, the antrum to
the animal's right of a line passing through *yz*, and a preantral part to the left of this line and extending to the line

---

[1] CANNON: American Journal of Physiology, 1898, I, p. 359. See also
ROUX and BATHAZARD: Comptes rendus de la soc. de biologie, 1897,
IV, p. 785; Archives de Physiologie, 1898, X, p. 85. A review of the
older literature on the movements of the stomach may be found in
CANNON's paper. See also BEAUMONT: Physiology of Digestion, Bur-
lington, 1847, p. 104; HOFMEISTER and SCHÜTZ: Archiv f. exper. Pathologie und Pharmakologie, 1885, XX, p. 1; ROSSBACH: Deutsches
Archiv f. klin. Medizin, 1890, XLVI, p. 296; HIRSCH: Centralblatt f.
klin. Medizin, 1892, XIII, p. 994.

through *wx*. The antrum is closed by the pyloric sphincter at *p*.

The wall of the stomach consists of three layers of smooth muscle fibres, an outer longitudinal, a middle circular, and an inner oblique coat. Smooth muscle is characterized physiologically by its power of slow rhythmic contraction and relaxation and its power of prolonged contraction (tonicity). The rôle that these characteristics play in determining the normal movements of the stomach can be best appre-

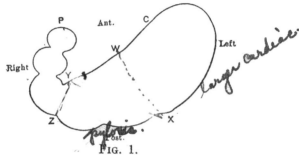

Fig. 1.

(Copied from CANNON: American Journal of Physiology, 1898, I, p. 360.)

ciated by a study of the accompanying figures, which indicate the changes that occur in the shape of the stomach after an ordinary meal (Fig. 2).

Within five minutes after a meal of bread a slight annular constriction appears near the duodenal end of the antrum and moves peristaltically towards the pylorus. This is followed by several other waves of similar character. Two or three minutes after the first movement is seen, very slight constrictions appear near the middle of the stomach (the preantral part) and becoming deeper move slowly toward the pylorus. As digestion goes on the antrum becomes somewhat elongated and the constrictions somewhat deeper, but never until the stomach is nearly empty do they divide the cavity entirely. The waves recur at intervals of almost exactly ten seconds and take about thirty-six seconds to pass from the middle of the stomach to the pylorus. When one wave is just beginning several others are therefore already

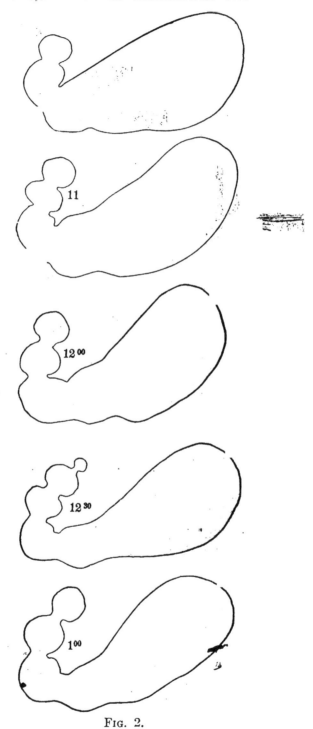

FIG. 2.

(Copied from CANNON: American Journal of Physiology, 1898, **I, p. 370.**)

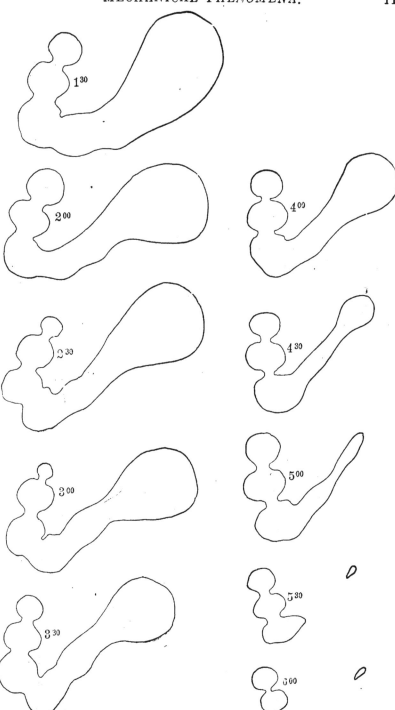

FIG. 2—*Continued.*

running before it as indicated in Fig. 2 (11.30–1.30). Between the rings of constriction the stomach is bulged out. CANNON has calculated the number of waves which pass over the stomach during a single digestive period lasting approximately seven hours as 2600.

The food slowly passes out of the stomach into the duodenum. The exact manner in which this happens has, however, been variously described by different authors. Those who with HIRSCH saw the food pass from the stomach into the duodenum at intervals are probably correct. In cats, CANNON found that no food appeared in the duodenum until the constrictions had been passing over the antrum for ten or fifteen minutes. When the food did appear it was squirted through the pylorus for some distance along the intestine. Every constriction wave does not force food through the pylorus. Several waves usually pass over the antrum before one is effective in this particular. At times two or three succeeding waves may each force food through the pylorus, but usually it remains closed for some time after it has allowed one constriction wave to pass food into the duodenum. The cause of this opening and closing of the pylorus will be discussed further on.

When a hard bit of food reaches the pylorus the sphincter closes tightly and remains closed longer than when the food is soft. This can be shown experimentally by feeding along with an ordinary meal pellets of bismuth subnitrate made up with starch paste. These can be readily recognized as darker spots in the general shadow cast by the gastric contents upon the fluoroscopic screen. When such pellets are given with the regular meal the stomach is emptied more slowly than when the food has a uniform consistency.

We have thus far spoken only of the movements of the pyloric half of the stomach. How does the cardiac half behave? For many years this has been looked upon as a sort of reservoir for the swallowed food, but it has always been considered purely passive, CANNON's experiments show that it is indeed a reservoir, but a most active one. The changes which the cardiac half of the stomach suffers during an ordinary digestive period is shown in the drawings, which were made by tracing the outlines of the stomach on tissue-paper laid over the abdomen of the cat at various times after feeding.

'By comparing the figures it can be seen that as digestion goes on the antrum seems to elongate and acquire a greater capacity, and that the constrictions make deeper indentations into it (Fig. 2, 11.00–1.30). When the fundus has lost most of its contents the longitudinal and circular fibres of the antrum contract and make it shorter and of less capacity once more. As compared with the changes in the form of the rest of the stomach those in the antrum are slight.

The first region to decrease markedly in size is the preantral part, at the beginning of which the peristaltic waves commence. These gradually force some of the stomach contents in this region into the antrum, so that the preantral part little by little begins to assume a tubular form in consequence of the sustained (tonic) contraction of the muscle fibres of this region (Fig. 2, 1.30–2.30). At one end of this tube we have the rounded fundus, at the other the actively contracting antrum. Shallow constrictions may pass along the tubular portion.

The muscle fibres (longitudinal, circular, and oblique) found in the fundus gradually contract upon the spherical mass of food found here and slowly force it into the tubular portion. The size of the fundus thus gradually diminishes, until the shadow cast by this portion of the stomach entirely disappears (Fig. 2, 5.00–5.30). The tubular portion of the stomach forces the food on into the antrum, until finally, when the fundus is empty, the last remnants of food are squeezed out of the tubular portion into the antrum.

We see from the above that the time which the food spends in the stomach is considerably longer than is commonly supposed. In the illustrations it can be seen that all the food had not passed out of the stomach seven hours after feeding. The same holds true for the human being, though absolute quantity of food and its chemical and physical character have much to do with its passage into the duodenum. A large meal would, other things being equal, take longer to leave the stomach than a smaller one. It was pointed out above that coarse particles of food delay the opening of the pylorus, and so keep a meal in the stomach a correspondingly longer time. One of the pernicious results of incomplete mastication of the food may well be traced to this fact. We shall see below how the opening and closing of the pyloric sphincter is affected still more powerfully by the chemical constitution of the food.

We can readily appreciate the value of an organ which, as the stomach, retains the swallowed food and only little by little passes it on into the intestinal canal beyond. In this way the food does not become heaped up in any section of the small or large bowel until the rectum is reached, and greater chance for the chemical elaboration of the various foodstuffs and for their absorption is obtained in consequence.

Having considered the movements of the stomach-wall, we must discuss briefly the movements of the food within the stomach. The older observations regarding this point are very contradictory.

As was shown above, waves pass rhythmically over the antrum. The food squeezed forward by an undulation may have one of two things happen to it. If the pylorus is open the wave serves to push the food on into the duodenum. We saw above, however, that by no means every wave is effective in this direction. For the majority of waves it might almost be·said the pylorus remains closed. Under these circumstances the food is forced into the blind, pouch-like extremity of the antrum. When this occurs a part at least of the food which is being pressed upon is forced backward through the constriction towards the cardiac end of the stomach. CANNON was able to show that the food actually moves in this way by mixing with it starch-paste pellets of bismuth subnitrate, the excursions of which could readily be followed among the other food. When the pylorus is clósed the food is squirted back through the oncoming constriction with considerable force. A subsequent wave then carries the food toward the pylorus once more. In this way the food is brought in contact with the mucous membrane of the stomach over and over again, and thus besides undergoing a certain degree of mechanical division becomes thoroughly mixed with the gastric juice. Interestingly enough, it is from this more active pyloric half of the stomach that the largest secretion of gastric juice occurs.

While digestion is going on and all the time that the antrum is most busily engaged in kneading and rekneading the food in this portion of the stomach the food in the cardiac half shows no sign of movement. Bismuth-subnitrate balls contained in the food which lies in the fundus keep their relative positions until the fundus begins to contract and then move slowly forward toward the antrum. This observation should settle for all time the question of salivary digestion in the stomach. As is well known[1] the amylase and maltase of the saliva do not act upon starch or maltose respectively when even a small percent of any acid is present. Careful examination of the fundus contents after a starchy meal by CANNON and DAY[2] have shown that no inconsiderable amount of salivary digestion occurs in the stomach. Herein we find another fact indicative of the importance of thorough mastication and insalivation of the food before it is swallowed. Food thus prepared can undergo salivary digestion in the cardiac half of the stomach, perhaps even for hours before it is brought to a standstill by becoming mixed with the hydrochloric acid of the gastric secretion.

---

[1] See p. 103.
[2] CANNON and DAY: American Journal of Physiology.

5. **The Passage of Different Foodstuffs from the Stomach. The Opening and Closing of the Pyloric Sphincter.**—It is the function of the pylorus, through the contraction of the muscular fibres contained in it, to keep the gastric contents from being forced out of the stomach while subjected to the churning movements of this organ. At certain times, however, the pylorus relaxes and allows a part of the gastric contents to pass through. Some of the older observers believed that the pylorus was contracted during the entire digestive period and that only when the food had been thoroughly mixed with the gastric juice did it relax and allow the stomach to empty itself entirely and at once. We know now, from the findings of EWALD, BOAS,[1] and PENZOLDT[2] with the stomach-tube, and the more perfect experiments of CANNON[3] with the x-rays, that the stomach empties itself little by little through periodic openings and closings of the pylorus. As has been shown above, the peristaltic waves of the stomach pass continuously under normal conditions from the middle of the stomach to the pylorus. As long as the pylorus is closed these waves serve only to churn the food and mix it thoroughly with the gastric secretion. When the pylorus relaxes, however, these same waves serve to push the gastric contents into the duodenum. What now determines this periodic opening and closing of the sphincter?

HIRSCH[4] observed in 1893 that when an acid is introduced into the duodenum through a duodenal fistula the escape of the gastric contents from the stomach is delayed for some time. SERDJUKOW[5] confirmed this finding in 1899. About the time of HIRSCH's observations, PENZOLDT found

---

[1] EWALD and BOAS: Virchow's Archiv, 1885, CI, p. 364.

[2] PENZOLDT. Deut. Arch. f. klin. Med., 1893, LI, p. 545.

[3] CANNON: Am. Jour. of Physiol., 1898, I, p. 368.

[4] HIRSCH: Centralblatt für klin. Medizin, 1893, XIV, p. 73.

[5] SERDJUKOW: Russian dissertation reviewed in Hermann's Jahresbericht ub. d. Fortschritte d. Physiologie, 1899, VIII, p. 214.

that those foods which delay the appearance of free hydrochloric acid in the stomach remain longest in this organ.
More recently CANNON [1] has reinvestigated this subject,
confirmed the findings of PENZOLDT, and outlined a theory
of the action of the pylorus which agrees with experimental
and clinical facts as we know them to-day.

CANNON investigated the rate at which different foodstuffs leave the stomach to enter the small intestine. As
examples of a nearly pure protein diet, boiled beef free from
fat, boiled whitefish or the white meat of fowls was used.
Beef-suet, mutton-fat or pork-fat served as nearly pure
fats, while starch paste, rice, and potatoes were taken as
examples of a carbohydrate diet. Definite amounts of the
various foods mixed with bismuth subnitrate were fed
to full-grown cats which had been without food for twenty-
four hours previously, and by means of the x-ray the
rapidity was noted with which the various foodstuffs escape
into the intestine. The time at which the food begins to
move into the duodenum can be accurately determined in
this way, and by measuring the aggregate length of the
shadows in the small intestine at half-hour or hourly intervals the relative amounts of food in the intestine from
time to time can be fairly well gauged.

In the following curves (Fig. 3) constructed from CAN-
NON's figures are indicated the different velocities with
which protein, fat, and carbohydrate leave the stomach.
It will be seen that the fats and carbohydrates begin to
move out of the stomach soon after ingestion, the carbohydrates leaving very rapidly, while the fats leave only
slowly. The curve representing the carbohydrates (curve C)
rises rapidly to a maximum which is reached at the end
of the second hour, to fall more slowly after this point is
passed. The curve for fats (curve A) both rises and falls
slowly, does not reach its maximum until the third hour,

---

[1] CANNON: Journal of the Am. Med. Assoc., 1905, XLIV, p. 15.

and at no time even approaches the maximum points reached by the carbohydrates or proteins. The protein curve (curve *B*) possesses the greatest interest. Not only do the proteins begin to pass out of the stomach much later than either fats or carbohydrates (often none of the gas-

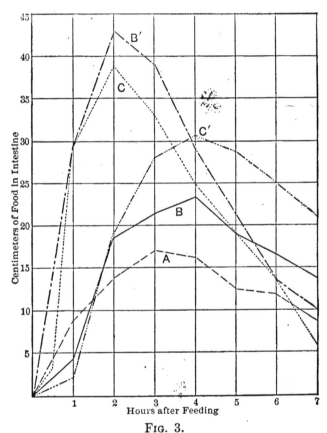

FIG. 3.

A =fat; *B* =protein; *B'* =acidulated protein; *C* =carbohydrate; *C'* =alkalinized carbohydrate.

tric contents leave the stomach before the end of the first hour after a protein meal), but their maximum velocity is not attained until the end of the fourth hour. It will be noticed that these findings harmonize with those of PENZOLDT.

What now determines that carbohydrates upon which

the gastric juice has no effect should leave the stomach quickly while proteins which are digested in the stomach should remain here a long time? The observation of PEN-ZOLDT that those foods which combine with the hydrochloric acid of the stomach are the last to leave this viscus has been pointed out above. (It is the presence of free hydrochloric acid in the stomach and near the pylorus that determines the relaxation of the sphincter and explains why different foodstuffs enter the small intestine at different rates.) . As will be shown later [1] an abundance of gastric juice is poured out on both a protein and a carbohydrate diet. A diet consisting chiefly of fat causes the secretion of much less juice. Now since carbohydrates cause a great secretion of gastric juice but do not unite chemically with the acid, free hydrochloric acid accumulates almost at once in the stomach. Proteins, on the other hand, unite with the acid of the gastric juice and hence prevent the accumulation of free acid for a considerable length of time. Finally, fat calls forth only a slight secretion of gastric juice, but that which is produced soon accumulates in the stomach, as it does not unite with the fat. These facts explain the different rates at which the various foods pass out of the stomach.

The idea that it is the presence of free hydrochloric acid in the stomach which determines the relaxation of the sphincter can be still further tested. If a carbohydrate meal is mixed with an alkaline fluid it delays the appearance of free hydrochloric acid in the stomach. Such a meal we find also leaves the stomach much later than the pure carbohydrate. This is indicated in curve C' in Fig. 3. It will be seen that the curve for alkalinized carbohydrate tends to approximate that for ordinary protein.

It is possible to try the converse of this experiment by feeding acidulated protein from which any excess of acid

---

[1] See Chapter XI, Part 2.

has been removed and noting the rate at which this food is discharged from the stomach. When acidulated protein is fed, the hydrochloric acid of the gastric juice is allowed to accumulate from the beginning. As is indicated in curve B' in Fig. 3, corresponding to this fact, we find that acidulated protein leaves the stomach as rapidly as pure carbohydrate,—in fact, the two curves almost coincide.

It might be thought at first that the addition of an acid to a pure carbohydrate meal would hasten its discharge from the stomach. Experiment shows that this is not the case, for a curve showing the velocity with which a meal of crackers mixed with 0.4 percent hydrochloric acid leaves the stomach practically coincides with that furnished by an equally large meal consisting of crackers and water only. How is this fact to be explained? We find an answer in the observations of HIRSCH and SERDJUKOW quoted above, who found that the presence of an acid in the *duodenum* causes the discharge of food from the stomach to be delayed. Whether this was due to a contraction of the pyloric sphincter or to a cessation of the movements of the stomach could, however, not be determined in their experiments. We know now from CANNON's researches that the peristaltic movements of the stomach are continuous, so that HIRSCH's and SERDJUKOW's findings must be explained through a contraction of the pyloric sphincter.

We see therefore that two factors are concerned in the opening and closing of the pylorus. The pylorus opens whenever free hydrochloric acid of sufficient concentration is present in the stomach. The opening of the pylorus allows the escape of a part of the acid stomach contents into the duodenum. As soon as the acid comes in contact with this portion of the intestinal tract, however, the pylorus is made to close and remains closed until the acid in the duodenum is neutralized through the flow of the pancreatic juice and bile into this portion of the gut. As will be shown later, the presence of acid in the duodenum is a determining condition

for the flow of juice from the pancreas.[1] As the acid in the
duodenum becomes neutralized the stimulus to the closure
of the pylorus is weakened until the acid in the stomach
once more opens the sphincter. Another portion of food
in consequence escapes from the stomach, the pylorus closes
once more, and the cycle is repeated.

Automatically, therefore, carbohydrates which are not
acted upon by the gastric juice leave the stomach soon after
ingestion and quickly, while proteins which are digested
in this viscus are retained and discharged only slowly. The
intestine is spared in this way large doses of acid stomach
contents which unless neutralized interfere so markedly
with the digestive functions of the intestine.

The behavior of fat differs a little from that of the carbohydrates and proteins. The immediate but slower discharge
from the stomach is explained in part by the small amount
of gastric juice which is poured out upon fat. The gastric
peristalsis after a meal consisting mainly of fat is also not
as vigorous as that found after a meal of protein or carbohydrate. Moreover, as soon as fats enter the duodenum they
cause the same closure of the pylorus which is caused by
acids.[2] In this way the fats are kept for a long time in the
stomach.

---

[1] See Chapter XII, Parts 2, 4 and 5.

[2] LINTWAREW: Biochemisches Centralblatt, 1903, I, p. 96, quoted
by CANNON.

CHAPTER II.

THE MECHANICAL PHENOMENA OF ALIMENTATION
(Continued).

6. The Movements of the Small Intestine.—A review of the literature on the movements of the intestine, both large and small, may be found in the articles of GRÜTZNER [1] and CANNON [2] and in the section on the intestine by STARLING [3] in SCHAEFER'S Text-book of Physiology. The following paragraphs sum up CANNON'S conclusions on the movements of the intestine as studied by the already outlined method of x-ray examination of the alimentary tract after a meal mixed with bismuth subnitrate. Cats fed on canned salmon or bread-and-milk mush were used for experimental study.

It is best to begin a description of the normal movements of the small intestine by referring to Fig. 4, in which is shown the appearance of the food in the intestine of a cat five and three-quarter-hours-after a meal of canned salmon. The animal is lying upon her back. In the middle of the plate in dim outline is seen the spinal cord with the pelvis below. On the right side above is the pyloric extremity of the stomach, and below it in dark shadow several intestinal loops lying over each other. In lighter outline and occupying the entire abdominal cavity are other loops also filled with food. The small dark shadow on the left is the cæcum.

---

[1] GRÜTZNER: Pflüger's Archiv, 1898, LXXI, p. 515.

[2] CANNON: American Journal of Physiology, 1902, VI, p. 251.

[3] STARLING: Schaefer's Text-book of Physiology, Edinburgh and London, 1900, II, p. 330.

22

FIG. 4.
(From an *x*-ray photograph kindly sent me by Dr. CANNON.)

23

When sufficient time has elapsed so that the food has been distributed through the intestine as indicated in Fig. 4, and the animal is exposed to the x-rays, it is found that most of the loops are in a state of perfect rest. This is due only to the excited state of the animal. After a time, if the cat is allowed to remain quiet, the intestinal movements begin. Two chief kinds of intestinal movement may then be recognized: first, movements of *rhythmic segmentation*, and secondly, *peristaltic* movements. By far the more common of these two movements is the first named. *Swaying movements*, such as have been described by the older observers, are also seen at times.

The character of the *rhythmic segmentation* can be best understood by reference to Fig. 5. The movements consist essentially in the sudden division of one of the long narrow strings of food lying in one of the loops shown in Fig. 4 into a large number of segments of nearly equal size, These segments are then suddenly divided and neighboring halves unite to form new segments, after which the process is repeated. This division and redivision of the food follow each other many times in succession. In line 1 of Fig. 5 is shown the food as it lies in a loop of intestine in the quiescent state. Suddenly as the movements of *rhythmic segmentation* evidence themselves, this string of food is divided into a large number of nearly equal segments as indicated in line 2. A moment later each of these segments is divided in two pieces, and the neighboring pieces unite to form a new segment. Thus a and b of line 2 unite to form the segment c of line 3. When this division occurs the end segments A and B remain small. But the next segmentation which divides the segments of line 3 restores the old order of things, for the end pieces reunite with the halves of their adjoining segments to form the new segments indicated in line 4. This series of segments it will be seen corresponds to the series shown in line 2. The process of division and redivision of the food continues under ordinary circumstances uninterruptedly for more than half an hour, during all of which time the form scarcely changes its position in the abdomen. In one instance the segmentation was seen to continue with only three short periods of inactivity for two hours and twenty minutes.

The above description holds for the usual movements of the small intestine, and when the food is not too thickly crowded in that region of the gut. When too large an amount of food is present in any given portion of the in-

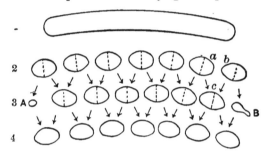

FIG. 5.

(Copied from CANNON: American Journal of Physiology, 1902, VI, p. 256.)

testine the constrictions may not completely divide the string of food, but only in part. This condition of affairs is illustrated in line 1 of Fig. 6.

Furthermore, the constrictions do not always take place in the middle of a segment, but more toward one side of it, as indicated by the dotted divisions in line 1 of Fig. 6. In this way one-third of each segment may be pinched off at each division, and only every third segmentation restores the original order of the series. Thus, in Fig. 6, the first division occurs along the dotted line *a* in each segment. This brings about the state of affairs shown in line 2. In this series the fragment of food at the extreme right of the string consists of one-third of an original segment, while that at the extreme left consists of about two-thirds of a segment. The next division which takes place along the dotted line *b* brings about the condition of affairs indicated in line

3, in which two-thirds of an original segment is present on the right-hand end of the string of food and one-third on the left hand. Not until the third division occurs along the dotted line *c* is the original arrangement of the segments restored.

The rapidity with which the segmenting movements occur is exceedingly interesting. If each segmentation, that is to say, if every change from one line to another in the diagrams, is counted, it is found that from twenty-eight to thirty occur each minute. If the string of food is thin, only rarely does the rate fall as low as twenty-three in the same unit of time. When much food is present in a given segment of intestine,

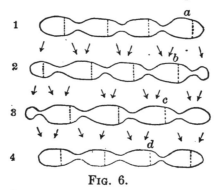

FIG. 6.

(Copied from CANNON: American Journal of Physiology, 1902, VI, p. 258.)

as indicated in Fig. 6, the number may fall as low as eighteen to twenty-one per minute.

CANNON has calculated that a slender string of food may commonly undergo division into small segments more than a thousand times without changing its position in the intestine. The admirable purpose which this serves in thoroughly mixing the food with the digestive juices and bringing it in contact with the absorbing mucous lining of the intestine will at once become apparent. Moreover, as MALL [1] has pointed out these repeated rhythmical contractions greatly

---

[1] MALL: Johns Hopkins Hospital Reports, I, p. 37.

aid in the propulsion of the blood into the portal circulation and the movement of the lymph through the lacteals. As absorption and secretion by the intestinal mucosa are both markedly influenced by the character of the blood which flows through this tissue, it can at once be seen how important these rhythmical contractions are.

We turn now to a consideration of the *peristaltic movements* in the small intestine. Peristalsis shows itself in two forms, first as a slow advancement of the food for a short distance in a coil, and secondly, as a rapid movement which sweeps the food through several loops of the gut. The latter is frequently seen under normal conditions when the food is carried forward from the duodenum. It is produced artificially by injecting soap-suds into the rectum.

When a mass of food has been subjected to the already described segmenting movements for a time, the latter may suddenly cease and the separate segments begin to move slowly forward, each segment following closely upon its predecessor. After moving forward in this way for a few centimeters the anterior segment comes to a stop, and the succeeding ones are swept into it, until the food lies stretched along the intestine as a solid, resting string of food.

Sometimes a single large mass of food, such as is indicated in Fig. 7, is pushed forward. A long string of food is first crowded together into a more rounded mass, such as is shown in line 1. Suddenly this mass is indented in the middle and assumes in consequence the shape shown in line 2. A second division now occurs in the posterior portion *a* which may cause the severed part to fly backward for some

FIG. 7.
(Copied from CANNON: American Journal of Physiology, 1902, VI p. 260.)

distance in the intestine, when a succeeding contraction again unites all the pieces into a rounded mass and pushes the whole slightly forward. This is shown in line 4 of Fig. 7.

7. The Movements of the Large Intestine.—As will be described in greater detail later the food passes from the small intestine through the ileocæcal valve into the ascending colon, and under normal circumstances this valve is competent to the food which has passed through it. From differences in physiological function we must distinguish between the first portion of the large gut which is composed of the ascending and transverse colon, and the second portion which is made up of the descending colon. Not only do the intestinal movements differ in these two portions of the large bowel, but also their contents. For while palpation shows that the food in the cæcum, ascending and transverse colons is soft, so that the walls of the gut can readily be approximated, it is found that the contents of the descending colon, sigmoid flexure, and rectum are made up of hard, incompressible lumps.

By far the commonest normal movement of the ascending and transverse portions of the large bowel is that of *anti-peristalsis*, that is to say, the peristaltic waves visible in this portion of the intestinal tract occur in a direction toward the stomach and away from the rectum. The first food which enters the colon from the small intestine is carried by these anti-peristaltic waves into the cæcum. The contents of the colon are not, therefore, as is generally believed, carried forward toward the rectum by a slow peristalsis, but are instead pushed backward a large number of times by these anti-peristaltic waves. These anti-peristaltic waves begin at the most advanced portion of the food, or, if considerable is present, at the splenic flexure, from which they sweep backward toward the cæcum. The average duration of a period of anti-peristalsis is four to five minutes, and the periods recur every ten to twenty minutes. Between the periods the colon is quiet.

As the ileocæcal valve is competent under normal circumstances, the anti-peristaltic waves do not force the food out of the large intestine back into the small intestine.

The constrictions as they pass over the colon therefore force the food into a blind pouch. As the food does not burst through the cæcum while subjected to the ever-increasing pressure of the contractile ring it can only escape through the advancing ring. We have already become familiar with this phenomenon in the stomach when the peristaltic waves pass over it against a closed pylorus. In the colon, therefore, we have the food again subjected to a thorough mixing, and another opportunity is presented for absorption.

The movements characteristic of the *descending colon* are a series of *tonic constrictions which pass downwards* toward the rectum. As the food accumulates in the ascending colon it is at first confined to the region of anti-peristalsis. As more food enters the ascending colon the contents of the large bowel are forced over more and more into the transverse and descending portions. After a time, as the food approximates the splenic flexure a deep constriction appears which pinches off a globular mass from the main body of the food as indicated in *A*, Fig. 8. This constriction *persists*, in other words, is a tonic one. The globular mass now moves slowly forward toward the rectum, and as it passes onward down the descending colon new constrictions appear which again pinch off globular masses from the advancing food column in the transverse colon.

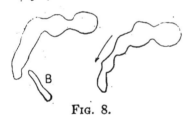

FIG. 8.

(Copied from CANNON: American Journal of Physiology, 1902, VI, p. 268.)

This is shown in Fig. 9. These rings slowly move down the descending colon, pushing the rounded masses of food before them until they reach the sigmoid flexure and rectum. Even during the short passage from the splenic flexure into the rectum some absorption occurs, for the food masses which are pinched off in the transverse colon are much softer than the dry fæces which collect in the rectum.

FIG. 9.
(From an *x*-ray photograph kindly sent me by Dr. CANNON.)

31

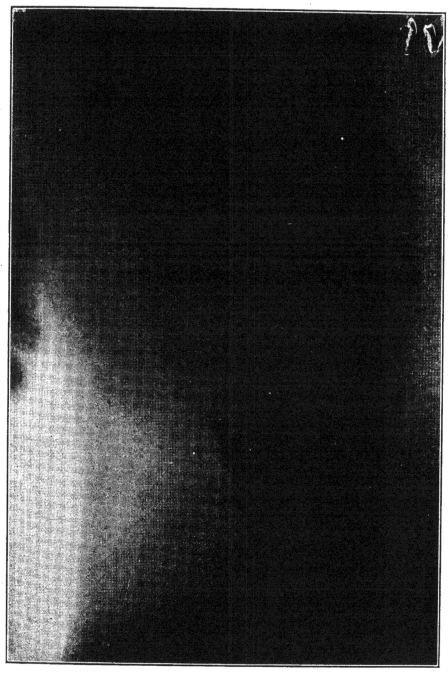

FIG. 10.
(From an x-ray photograph kindly sent me by Dr. CANNON.)

8. Defæcation.—The act of defæcation is in part an involuntary, and in part a voluntary, process. The first part of the act is involuntary and may be described as follows: In the cat the entire large bowel seems to swing around so that the ascending colon is raised to occupy in part the position formerly held by the transverse, and this the position originally occupied by the descending colon. This is apparent when Fig. 11, which shows the beginning of the act of defæcation, is compared with Fig. 10, which shows the position of the large intestine under ordinary circumstances. The tonic constrictions pictured in Fig. 10 disappear at the same time, and their place is taken by a single broad contraction of the circular muscle which tapers off the intestinal contents on both sides of it. In this way a fæcal mass lying low in the descending colon is pinched off from the intestinal contents higher up. The broad contraction now passes slowly downward and, aided by the voluntary contraction of the abdominal muscles and the voluntary relaxation of the anal sphincter, which constitute the second part of the act of defæcation, pushes the fæcal mass out of the canal.

FIG. 11.
(Copied from CANNON. American Journal of Physiology 1902, VI p. 270.)

When the external act of defæcation has been thus accomplished, the colon with its remaining contents returns to nearly its former position (Fig. 12). From the time that the colon begins to change its position until it returns to it once more takes about 20 minutes. When the original position has been reassumed the slowly moving contractions begin again, and the contents of the large intestine which were not lost in the act of defæcation are slowly spread into the emptied portion of the descending colon.

Just how the intestinal contents get from the region of the

anti-peristaltic waves in the colon to the region of the slowly advancing rings is not yet entirely understood. In part the later portions of the food push that which has gone before ahead of it. But the ascending and transverse portions of the colon can get rid of most of their contents even in starvation, though they are, perhaps, never entirely emptied even under these conditions. Apparently strong tonic *peristaltic* waves at times pass over the ascending and transverse colon and force the contents of these portions of the large bowel into the region of the descending colon, where peristaltic waves are the rule.

9. **The Movement of Food through the Alimentary Tract as a Whole.**—The process of mastication takes a varying length of time, depending upon the nature of the food and to a large extent upon the individual. Dry food takes longest to become mixed with saliva sufficiently to allow of its being swallowed, while liquids are at once passed backward toward the œsophagus. The entire length of time required for the food to fall into the stomach after mastication is completed varies with different animals, the extremes being represented by four and twelve seconds. In man from four to seven seconds are required from the initiation of the act of deglutition to the time the food falls into the stomach,

The food remains in the stomach a variable number of hours, depending upon the quantity and the quality of the food ingested. Other things being equal, a small amount of food will escape into the small intestine in shorter time than a larger amount. The time that food remains in the stomach is probably greater than is generally believed to be the case. An average meal does not ordinarily leave the stomach entirely in less than six hours. The quality of the food plays an important rôle by determining the frequency with which the pyloric sphincter relaxes and allows the food to be forced onward into the duodenum by the contractile waves which pass over the antrum. As was shown above, not every wave forces food out of the stomach.

FIG. 12.
(From an x-ray photograph kindly sent me by Dr. CANNON.)

37

When the pylorus relaxes the food is squirted for a considerable distance along the duodenum (Fig. 13), where it lies
quietly until added to by further contributions from the
stomach. In this way a long thin string of food is formed.
During all this time the pancreatic and bile ducts pour the
secretions from their respective organs into the food. All at
once the string of food breaks up into several segments, and
the process of rhythmic segmenta-
tion already described above is
started. After this has continued for
some minutes the segments unite
into a single large mass, or into
groups, and slowly pass along the
gut. Near the pylorus the peristalsis
is more rapid than lower down in the
small intestine. "The masses once

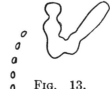

Fig. 13.
(Copied from Cannon:
American Journal of
Physiology, 1902, VI;
p. 262.)

started go flying along, turning curves, whisking hither and
thither in the loops, moving swiftly and continuously forward."[1] After passing forward in this way for some distance the food is collected again into thick and long strings
and the process of segmentation repeated. The strings
of food may remain lying quietly in a loop of intestine
for an hour or more. During the first stages of digestion the
food lies chiefly on the right side of the abdomen, during the
later stages chiefly on the left. By these combined movements of segmentation and peristaltic advance, both of which
are repeated from time to time, the food is finally brought to
the ileocæcal valve.

The time elapsing before the food enters the duodenum
from the stomach varies, as already pointed out, with
the nature of the food. In general it may be said that no
food enters the small intestine until one or one and a half
hours after eating. Five to six hours elapse after eating
before food begins to appear in the large intestine, so that it

---

[1] Cannon: American Journal of Physiology, 1902, VI, p. 263.

is evident that approximately five hours are required by the
food to traverse the small intestine.

The food about to pass into the large intestine is directed
toward the latter from some distance back in the small gut (see
Fig. 8, B). From here it moves slowly along the ileum and is
pushed through the valve into the colon to fall into the
cæcum. This passage of food through the ileocæcal valve
seems to act as a stimulus which excites the colon to activity.
As the food approaches the ileocæcal valve the large intestine
is quiet and relaxed. As soon, however, as the food has
entered it a strong contraction takes place along the cæcum
and lower portion of the ascending colon, which is followed
immediately by the anti-peristaltic waves which have already
been described and which continue running for two or three
minutes.

Under ordinary circumstances the succeeding masses of
food force the older portions onward through the large
bowel, but even in starvation most of the contents of the
large bowel are gotten rid of. But a *complete* emptying
of the large intestine seems never to occur. The time
which the food spends in the large intestine varies of course
with the intervals elapsing between succeeding defæcations.
The time which the intestinal contents spend in the ascending and transverse portions of the large bowel is certainly to
be measured in hours. During all this time absorption is
actively going on. As the food passes into the descending
colon, sigmoid flexure, and rectum this absorption is probably
considerably diminished.

The voluntary part of the act of defæcation is preceded
by an involuntary act of preparation which takes an hour
or more. This is the time required for the intestinal contents to pass from above into that portion of the lower
bowel which has been emptied by the last act of defæcation.
The voluntary part of the act of defæcation takes only a
few seconds, depending upon the amount and consistency
of the defæcated mass.

10. The Nervous Control of the Alimentary Tract.--We are still far from a correct understanding of the relation of the nervous system to the movements of the various parts of the alimentary tract. The experimental results obtained by the score of investigators who have busied themselves with this problem do anything but harmonize, a fact not strange when the difficulties standing in the way of the solution of the problem are considered. Narcotics, operative procedures, etc., all so markedly influence the movements of the alimentary tract that when these constitute the necessary means which must be utilized in a study of the problem uniform results can scarcely be expected. The following paragraphs follow in the main STARLING's[1] recent review of the subject.

The act of *deglutition* is only in part voluntary, and may as a whole be considered as an essentially reflex act. The reflex is initiated whenever the palatine branches of the trifacial, the glosso-pharyngeal, and superior laryngeal nerves are stimulated either through the presence of food or saliva in the mouth or by artificial means as when an electrical stimulus is applied to the central end of the divided superior laryngeal nerve. The afferent nerves pass into the medulla, in the upper portion of which is situated a "centre?" whose destruction is associated with impairment or total loss of the power to swallow. The impulses which go to bring about a movement of the muscles concerned in the act of swallowing leave the medulla chiefly by way of the trifacial, facial, and glosso-pharyngeal. The œsophagus is supplied almost solely by the vagus.

The œsophagus has the power of spontaneous peristaltic contractions, within the body, however, the waves which pass over the œsophagus in the ordinary act of swallowing seem to be intimately connected with an uninjured nervous system. Division of the nerves supplying the œsophagus

---

[1] STARLING: Ergebnisse der Physiologie, 1902, I, 2te Abth., p. 446.

causes a stoppage of the œsophageal movements. Simple ligature of the œsophagus without injury to the nerves does not, however, keep the œsophageal movements from passing over the entire length of the œsophagus. A piece can even be cut out of the œsophagus and if the nervous connections between the upper and lower ends have not been injured the swallowing movements inaugurated in the upper end pass (by way of the nerves) over the lower end also. It seems therefore as though the ordered act of deglutition which is started through stimulation of certain afferent nerves is dependent in the last analysis upon impulses which pass from an excited "centre" by way of certain efferent nerves to the œsophagus, one segment after another of which is thereby made to contract.

The *stomach* is supplied with cranial nerves by way of the two vagi, and with sympathetic nerve fibres by way of the splanchnics and the solar plexus. The vagi contain fibres which not only bring about movements in the stomach but also such as inhibit movement. Under ordinary circumstances, more especially when the stomach contains food, stimulation of the vagi brings about a contraction of the cardiac end of the stomach; or one or a series of contractions arise in the preantral portion of the stomach; or, finally, the musculature of the whole stomach may go into a state of tonic contraction which gradually increases and then after a longer or shorter period of sustained contraction as gradually relaxes. Under certain circumstances stimulation of the vagi may have an opposite effect. Especially after the administration of pilocarpin may stimulation be followed by muscular relaxation. The sympathetic nerves supplying the stomach in a certain sense antagonize the action of the vagi. If the splanchnic nerves are stimulated a decrease in the tonus of the gastric musculature as well as a decrease in the rhythmical contractions of the stomach are usually observed.

In addition to the nerves which run into the wall of the

stomach, there exist in this viscus isolated ganglion cells and nerve fibres from the plexus of AUERBACH and MEISSNER. These local nervous elements have been looked upon as causing the rhythmical contractions of the stomach, but it is questionable whether the unstriped muscle fibres are not themselves responsible for this. It approximates correctness most nearly, no doubt, when we say that the unstriped muscle fibres of the stomach are capable of the rhythmical and the sustained contractions which we observe in this organ, but that these contractions can be markedly influenced through the nervous system.

In addition to the local nerve-cells and plexuses, and the vagus and sympathetic fibres which go to the stomach, there exist in the spinal cord and the ganglia at the base of the brain so-called "centres" which on stimulation lead to muscular contractions or relaxations in the stomach, but we do not understand how these different elements coöperate to bring about the ordered movements observed in this viscus after an ordinary meal, or the disturbances noted in certain pathological states.

The *small intestine* is supplied by branches from the vagus nerves and from the sympathetic system. The sympathetic fibres come to the intestine in part from the solar and lumbar plexuses, in part by way of the splanchnics. Stimulation of the vagus brings about in the small intestine as in the stomach motor effects which may evidence themselves in rhythmical contractions, or in sustained tonic contractions. The chief effect of stimulation of the sympathetic fibres seems to be that of inhibition. As was first shown by PFLÜGER, stimulation of the splanchnics leads to cessation of movement in the small intestine. From MALL's [1] careful studies it is known that an anæmia of the intestines causes a cessation of movement in them, and it was once thought that the inhibition of movement when the splanch-

---

[1] MALL: Johns Hopkins Hospital Reports, 1896, I, p. 37.

nics are stimulated was secondary to the anæmia brought about by this means. But, as BAYLISS and STARLING have shown, such inhibition of intestinal movement still follows stimulation of the splanchnics in freshly killed animals, in other words, when no circulation is present. Nothing definite seems to be known regarding the rôle of the plexuses of AUERBACH and MEISSNER in the small intestine. Whether they are, in the last analysis, responsible for the rhythmical contractions of the jejunum and ileum seems questionable. This power probably resides in the musculature itself.

Stimulation of certain regions in the ganglia at the base of the brain and certain portions of the spinal cord leads to muscular contractions or relaxations in the small intestine by way of the vagus and sympathetic nerves supplying the gut. Some connection must also exist between the cerebrum and the small intestine, for, as the next paragraphs show, mental states may cause a stoppage of all movement.

It has long been a recognized clinical fact that *emotional states* such as fear, anxiety, sorrow, etc., bring in their train a long series of digestive disturbances. The physiological pathology of these disturbances has recently been put on a more scientific basis by a series of experimental observations. We shall later become acquainted with facts which indicate how largely certain secretory phenomena of the alimentary tract are influenced by psychic states. Here mention must be made of the important relation which exists between mental states and the movements of the stomach and intestine.[1]

CANNON observed in his studies of gastro-intestinal movements that female cats were much more suitable for experimental purposes than males. Even though treated exactly alike the movements which appeared almost with-

---

[1] CANNON: American Journal of Physiology, 1898, I, p. 380; ibid., 1902, VI, p. 260.

out exception in female cats were almost as constantly
wanting in male cats. Accompanying this was always a
difference in the behavior of the cats when bound into
the holder for observation. While the females would lie
quietly and purr, the males would fly into a rage and
struggle to get free. The following observation showed
the direct connection between the mental state and the
movements of the gastro-intestinal tract. A male cat
which had been fed an hour and a half previously was ·tied
into the holder. The waves were passing regularly over
the stomach at the rate of six a minute. This had not
lasted long when the cat fell into a rage and all movement
in the stomach ceased at once.

A similar relation exists between the mental state and
the movements of the intestine. This is not surprising
when it is remembered that the nerves which are distributed
to these two portions of the alimentary tract are the same.
Whenever a cat becomes enraged and for some time after
it is again pacified the movements of the small and large
intestine cease entirely. Even when an animal is only
slightly restless in the holder no intestinal movement may
be apparent. In a continuously
fretful cat this may continue for
an hour. Between the periods of
excitement the intestinal movements go on in a normal manner.
When the segmenting movements
in the small intestine cease, the
segments of food coalesce to form
a single long string. In Fig. 14,

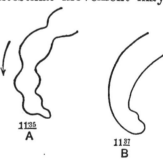

FIG. 14.

A, is shown the appearance of the (Copied from CANNON: Ameri-
large intestine when the anti-per- can Journal of Physiology,
istaltic waves are running nor- 1902, VI, p. 275.)
mally, and in Fig. 14, B, how the same region of the intestine
looks when the anti-peristaltic waves are inhibited through
excitement. Interesting is the fact that the tonic contrac-

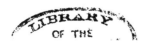

tions of the descending colon are apparently not affected by emotional states, for they do not relax in the excitement which causes the other movements to cease.

Inhibition of intestinal movement is not the only consequence of excited mental states   CANNON quotes the experiments of ESSELMONT and FUBINI to show this.   ESSELMONT found that in the dog signs of emotion always markedly *increase* the motor activities of the intestine, though only for a few moments.   FUBINI noted that fear brings about an increased peristalsis.

Finally, the fact must be mentioned that all movements of the intestines go on during sleep in the same way as in the waking hours.

11. On the Action of Saline Cathartics.—According to the generally accepted view those salts which are classed under the head of the saline cathartics in the works on pharmacology are believed to exert their action by preventing the absorption of water from the intestinal contents as the latter pass through the alimentary tract.   In this way it is believed that the ordinary inspissation into the compact fæces which collect in the lower bowel is prevented, and the intestinal tract is rid of its contents in the form of very soft or even liquid stools.

It seems from experiments recently carried out by J. B. MacCALLUM [1] that this conception can no longer be looked upon as the correct one.   MacCALLUM's experiments were performed chiefly on rabbits, but dogs and cats were also employed. The salts used were sodium citrate, sulphate, tartrate, oxalate, phosphate, and fluoride, barium chloride, and magnesium sulphate.   These experiments have shown that the saline purgatives act not only when introduced into the intestine, but also when injected subcutaneously or intravenously.   They act most powerfully, however, when applied directly to the peritoneal coat of the intestine.   When 5 to 10 c.c. of a one-

---

[1] MacCALLUM, J. B.: American Journal of Physiology, 1903, X, p. 101; Pflüger's Archiv, 1904, CIV, p. 421.

eighth molecular [1] sodium citrate solution are injected into the lumen of the intestine of a rabbit, increased peristaltic movements manifest themselves in 10 to 15 minutes. About the same length of time is required when the same amount of this salt solution is injected subcutaneously. When, however, only 1 to 2 c.c. of this sodium citrate solution are injected intravenously, a striking increase in intestinal movements is visible in one to two minutes. It almost seems from these experiments that the saline cathartics must be absorbed from the intestinal tract before they can produce their specific effects. A reaction in the form of a local constriction of the musculature of the gut is obtained almost immediately after painting one of the saline cathartics on the peritoneal coat of the intestine.

When equimolecular [2] solutions are compared, it is found that by far the most powerful of the cathartics listed above is barium chloride, after which come the citrate, fluoride, sulphate, tartrate, oxalate, and phosphate of sodium, the intensity of the action of which decreases approximately in the order named. The intravenous injection of a solution containing a few milligrams of the dry salt is sufficient to bring about powerful contractions of the intestine. The power of the various salts to produce their cathartic action was determined by discovering the lowest concentration in which they

---

[1] That is, a solution of sodium citrate made by dissolving one gram-molecule of the dry salt in enough water to make eight litres. A gram-molecule of a substance is the molecular weight of that substance (plus the molecular weight of its water of crystallization, if it has any) expressed in grams.

[2] That is, solutions containing the same number of gram-molecules of the various salts dissolved in the unit volume of the solvent (water in this case). The comparison of equal percentage solutions, that is solutions containing the same weight of the salts in the unit volume of solvent, as was generally done by the older observers, leads to entirely erroneous conceptions of the relative activity of the dissolved salts. It would be well if all workers in experimental medicine would employ only chemically equivalent solutions.

were effective. Magnesium sulphate is about as active a cathartic as sodium sulphate, but it is by no means as harmless as the latter. MacCallum found that magnesium sulphate was often fatal in doses in which sodium sulphate is entirely harmless. This statement is confirmed by some of my own experiments on glycosuria, in which it was found that the action of the sulphate of sodium is less harmful than that of magnesium. The poisonous action is apparently determined by the magnesium constituent in the latter salt, which has a powerful effect upon the heart. A practical conclusion to be drawn from this is that it is better to give the sodium salt to patients than the magnesium salt. The intensely poisonous action of the barium chloride should also put a ban upon the use of this drug in medicine. The fluoride and oxalate also have specific poisonous properties which speak against their use in medicine, especially when apparently harmless salts may be employed with just as good results if only care be taken to use the right amounts.

The purgative action of the saline cathartics depends upon yet another factor than the mere increase in the peristaltic movements of the intestine, namely, an increased secretion of fluid into the intestine. This fact was observed by Gumilewski [1] and Röhmann,[2] and is confirmed by MacCallum's observations. If a loop of intestine is tied off and a saline cathartic is injected subcutaneously or intravenously, or is simply painted on its surface, a secretion of fluid into the loop of intestine is observed in addition to the increased peristalsis.

MacCallum has been able to show that the effect of the saline cathartics enumerated above, both in bringing about an increased peristalsis and an increased secretion of fluid into the intestine, can be counteracted by calcium chloride, and to a less extent by strontium and magnesium chloride. This antagonism between sodium and calcium salts seems to

---

[1] Gumilewski: Pflüger's Archiv, 1886, XXXIX, p. 556,

[2] Röhmann: Pflüger's Archiv, 1887, XLI, p. 411.

exist in a variety of physiological reactions. RINGER [1] first called attention to it in his experiments on heart muscle. He found that the contractions of strips of this tissue, which are beating rhythmically in sodium chloride solutions, can be inhibited through the addition of a calcium salt to the sodium chloride solution. The experiments of LOEB, who has elaborated those of RINGER, show that the same antagonism exists in the case of voluntary muscles and in nerves. Experiments of other investigators are at hand which show that the antagonism between calcium and sodium salts exists in the involuntary muscles also, but these experiments are not entirely free from criticism, for no special means were taken to exclude the effects of nerve fibres or nerve-cells present in the preparations.

Whether the increased peristalsis is brought about directly through the action of the saline cathartics upon the muscle-cells themselves or only indirectly through an action of the salts upon the nerve plexuses of AUERBACH and MEISSNER has not as yet been definitely settled. It is certain that we do not need to go beyond the wall of the intestine for an explanation of the action of these drugs. This is proven by the fact that pieces of gut ligatured and cut out of the body will show their characteristic movements when immersed in solutions of the various salines. A secretion of fluid will even occur into such pieces of intestine. This indicates that the explanation of this phenomenon too does not lie beyond the walls of the alimentary tract. From the fact that very dilute solutions of the cathartics produce very violent and yet entirely local contractions in the intestine when painted upon its peritoneal surface it seems highly probable that the increased peristalsis is brought about through a direct action upon the muscular coat. If a stimulation of nerves lay at the foundation of the increased intestinal movements we would expect a more general effect from the local applications of the cathartic salts.

---

[1] RINGER: Journal of Physiology, 1884, V, p. 247.

So far as the quantitative relation is concerned which exists between the dose of a saline purgative sufficient to affect the intestine and that of calcium chloride sufficient to suppress the increased peristalsis and intestinal secretion brought about by the former, it may be said that a *chemically* equivalent amount of the one just counteracts the other. If, for example, a certain number of cubic centimeters of a 1/8 molecular sodium citrate solution are injected into a rabbit it requires an equal volume of a 1/8 molecular calcium chloride solution to counteract the effect of the former. This counteraction takes place almost immediately if the calcium chloride is applied to the peritoneal coat of the intestine. If the calcium chloride is injected intravenously it shows its specific effect in one to two minutes, but it takes ten to twenty minutes if it is injected subcutaneously or into the lumen of the intestine.

MacCallum is inclined to explain the action of the saline purgatives through their power of diminishing the concentration of the free calcium ions in the tissues upon which these cathartics act. This is the same explanation that Loeb gives of the twitchings observed by him when voluntary muscles are immersed in these same salt solutions. The addition of a calcium salt therefore counteracts the effect of the saline cathartics by restoring the concentration of the calcium ions and so bringing the intestine back to its original state. It must be remembered, however, that barium chloride is the most powerful of all the salts studied, and yet it is difficult to see how the administration of such traces of this substance as are necessary to bring about a violent catharsis is to be explained by its effects in reducing the concentration of the free calcium ions. Hofmeister's belief that catharsis is brought about through the coagulation of certain colloids by the cathartic salts seems less open to objection.

The movements of the intestine, *as a whole*, under the influence of saline cathartics have not yet been studied.

Whether they are merely exaggerated forms of the normal movements or differ materially from the latter cannot therefore be said with definiteness. That food takes a very much shorter time to pass the length of the alimentary tract under the influence of saline cathartics is a well-known clinical fact, and has often enough been proven experimentally by feeding inert substances and determining the time required before they appear in the fæces. The greater fluidity of the stools we must, in the face of these recent experiments, attribute to the increased secretion of fluid into the intestine and not to a decreased absorption. That less of the food is absorbed when saline cathartics are administered shortly after a meal is in part dependent, no doubt, upon the increased velocity with which the food traverses the bowel. Less time is allowed in consequence for the products of digestion to diffuse through the intestinal mucosa.

The cause of the pain (colic) which follows the ingestion of the saline (and other) cathartics is also not entirely understood. S. J. Meltzer has made an interesting study of the subject. This author holds the increased force of the contractions and their irregularity responsible for the pain in all forms of intestinal colic.

Further work with the x-rays would teach us much regarding the movements of the intestines as a whole under abnormal conditions. That after the administration of the saline cathartics it is probably a rapid sweeping peristalsis which advances the food through several coils of intestine, instead of the normal slow peristalsis which moves the food onward only a short distance, seems very probable from a statement made by Cannon,[1] who observed this form of movement in the small intestine as a regular consequence of the injection of soap-suds. Many of the soaps resemble in their action the saline cathartics.

12. **The Fate of Nutrient Enemas.**—While certain clinical

---

[1] Cannon: American Journal of Physiology, 1902, VI, p. 260.

observers have been unanimous in claiming that nutrient enemas introduced into the rectum are absorbed and utilized by the patient, others have been equally certain that they are of little or no value. The burden of evidence is, however, in favor of the good results which are obtained by this method of artificial feeding. Lack of success is no doubt in large measure due to mistakes made in the composition of the enemas used, for the absorption of many foods can occur from the intestinal tract only after they have been acted upon by the digestive juices.

Leaving aside for the moment the chemical aspects of the problem, two great objections have been brought against the efficacy of rectal feeding. The first of these is that the rectum has no powers of absorption, the second that the food does not pass from the rectum to a portion of the bowel, where it can be absorbed. It must be confessed that experimental evidence has until recently been largely in favor of these views. With scarcely an exception, investigators have come to the conclusion that food introduced into the rectum does not move far away from it. Against this idea has stood GRÜTZNER,[1] who found that when certain easily recognizable substances, such as starch grains, hair, or charcoal, suspended in normal salt solution are injected into the rectum they are carried upward through the bowel into the small intestine, even as far as the stomach. More recently CANNON[2] has brought direct proof of the movement of nutrient enemas from the rectum, not only through the large intestine but even into the small intestine. If we admit, therefore, that the rectum has but feeble or no powers of absorption, we can no longer maintain that the enemas. are not moved to a place in the intestine in which absorption is possible. The following may help to elucidate what has been said:

---

[1] GRÜTZNER: Deutsche med. Wochenschrift, 1894, XX, p. 897.
[2] CANNON: American Journal of Physiology, 1902, VI, p. 272.

In order to ascertain the fate of enemas, CANNON introduced various amounts of thick and thin food mixtures, holding bismuth subnitrate powder in suspension, into the previously cleaned rectum of various animals and observed what happened by means of the x-rays. The enemas had the following composition:

> 100 c.c. milk,
> 2 grams starch,
> 1 egg,
> 15 grams bismuth subnitrate.

To make the enemas thick all these were stirred together and boiled to a soft mush. To have them thin the egg was omitted until after boiling, when it was added to the cooled mass. The amounts injected varied from 25 c.c. to 90 c.c., which was about sufficient to fill the large bowel of the animals experimented upon.

Besides depending upon mere observation, radiographs were taken at various intervals, in order to show the distribution of the injected enemas. After a control radiograph had shown the absence of any bismuth-containing food in the intestine of the animal to be experimented upon, the food was introduced into the rectum. It was found in these experiments that when only small amounts are injected they lie at first in the descending colon. In every case, however, anti-peristaltic waves commence and the food is carried through the transverse and ascending colon into the cæcum. Small injections of nutrient material never pass this point. The larger injections, however, do not stop when they reach the ileocæcal valve, but pass through it high up into the small intestine. Strange as it may seem, this valve, which is competent to the food passing through it in the normal progress of the food from the stomach to the rectum, allows the nutrient material from large rectal enemas to pass through into the small intestine. The anti-peristaltic waves of the ascending colon seem to be the effective agents in forcing the enema backward through the ileocæcal valve and along the ileum; for when the valve first allows the food to pass through, it pours into the small intestine and appears as a mass which suddenly fills several loops of the gut. Anti-peristalsis does not appear in the small intestine though it continues in the large. After the food has been in the small intestine for some time, the typical segmenting movements .a$_p$p$_e$a$_r$, which divide and redivide the food in the manner already described above.

Figs. 15, 16, and 17 are radiographs which indicate how the food is forced after a large enema more and more from the large intestine into the small. An enema of 90 c.c. was given at 1.40 P.M. The first radiograph shows how at 1.50 P.M. the food is evenly distributed through the large intestine and through several loops of small intestine. In the second and third radiographs, taken at 2.15 and 3.00 P.M. respectively, the food is seen to have left the large intestine to a great extent, and to be occupying an increasingly greater number of loops of small intestine. Observations made at 3.00 P.M. showed segmentation to be going on in many of the loops.

The importance of the ascending and transverse portions of the large intestine as an absorptive organ has already been pointed out, and it is not surprising, therefore, to find evidences of this same absorption when food is introduced into the intestinal tract by way of the rectum. In the region of the anti-peristaltic waves in the large intestine the shadows become progressively lighter, until in the end the bismuth seems to be covering only the walls of the gut. In the descending colon the shadows retain their original opacity. When the rectal injections are large, the small intestine also comes into play as an absorptive organ, and this in the same way apparently as when the food is introduced through the mouth.

It must be pointed out, finally, that if any digestive juices are present in the large and small intestines, these are of course mixed with the nutrient enemas and so are given an

Fig. 15.—1.50 p.m.

-ray photograph kindly sent me by Dr. Cannon.)

Fig. 16.—2.15 P.M.
(From an x-ray photograph kindly sent me by Dr. CANNON.)

Fig. 17.—3.00 P.M

(From an x-ray photograph kindly sent me by Dr. Cannon.)

opportunity of exerting their specific effects. How much of such a secretion occurs, and its quality, can, however, only be surmised. If no food has been introduced through the mouth, and no "psychic" or other secretion from the stomach has in consequence been called forth, little or no pancreatic or intestinal juice of any kind may be present in the intestine. This urges upon us the necessity of introducing into the rectum only those foods which can be taken up directly by the lower portions of the alimentary tract. Certain foods are already in this condition, while others must first be digested outside of the body, or have introduced along with them those ferments which are found in the body and are capable of so acting upon the various constituents of the nutrient enemas that they are converted into readily absorbable substances. Unless this chemical side of rectal feeding is considered, we must fail in the objects which we wish to attain by these artificial means.

CHAPTER III.

THE JUICES POURED OUT UPON THE FOOD AND THEIR CHEMICAL CONSTITUENTS.

1. The Saliva.—This is the name applied to the mixed secretions from the salivary glands and the mucous glands of the mouth. The salivary glands are three in number, the parotid, submaxillary, and sublingual, and as their secretions differ somewhat they are best considered separately.

Human *parotid* saliva can be obtained by introducing a fine cannula into STENSON's duct. In this way a clear, watery liquid mixed with some epithelial cells and occasionally a few leucocytes is obtained. The reaction is usually given as faintly alkaline, but this seems to be incorrect. It is probably neutral, perhaps even slightly acid.

The amount secreted in twenty-four hours varies within wide limits physiologically. OHL gives 80 to 100 c.c. The specific gravity varies under physiological conditions and is stated to be between 1.006 and 1.012. The parotid secretion contains mucin and only traces of proteins. It contains the starch-splitting ferment *amylase* and the maltose-splitting ferment *maltase*. It also contains sulphocyanic acid in combination with the metals found universally in the tissue fluids. The exact amount of each of these substances varies under physiological conditions, so that it is not surprising that the figures given by different investigators do not agree very well. MITSCHERLICH found 14 to 16 parts, HOPPE-SEYLER about 7 parts of solids in 1000 of parotid saliva. About half of this amount is organic, the rest inorganic.

62

Potassium sulphocyanate is present to the extent of about 3 parts in 10,000.[1]

Human *submaxillary* saliva may be obtained in a way similar to that described above for obtaining parotid saliva, namely, insertion of a fine cannula into WHARTON's duct. When the sublingual gland does not open into the mouth by a separate duct, but discharges its secretion into WHARTON's duct, the two secretions can be separated by pushing the cannula far enough along the common duct to pass beyond the opening from the sublingual gland.[2] The secretion obtained in this way is clear and watery and of a specific gravity somewhat lower than that of the parotid. It is variously stated to be from 1.0026 to 1.0250, and is subject to considerable variation under physiological conditions. The amount of solids contained in the submaxillary saliva is between 4 and 5 parts in 1000, against 6 to 15 parts found in parotid saliva. The character of the solid constituents is much the same as in the parotid saliva, except sulphocyanate, the presence of which is questioned by some authors. Sulphocyanate is certainly present in less amount in the submaxillary secretion than in the parotid. This amount as well as the quantity of submaxillary saliva secreted in any unit of time varies greatly under physiological conditions. Roughly speaking, the total amount secreted in twenty-four hours is three times as great as that secreted by the parotid (OHL).[3] The reaction is generally stated to be alkaline, more probably, however, it is neutral. The starch-splitting activity of the submaxillary saliva is very great.

OHL seems to be the only investigator who ever obtained human *sublingual* saliva in sufficient amount to determine that it contains amylase, sulphocyanate, and mucin.

[1] VIERORDT: Tabellen, Jena, 1888, p. 30.

[2] See MOORE: Text-book of Physiology, edited by SCHAEFER, Edinburgh, 1898, I, p. 342.

[3] VIERORDT: Tabellen, Jena, 1888, p. 130.

. The mucous glands of the mouth secrete a small amount of an exceedingly tenacious material containing much mucin. This secretion has never been obtained pure in the human being, but experimentally it has been gotten from dogs after ligature of the ducts or extirpation of the entire set of salivary glands. The secretion is mixed with many epithelial cells from the mouth.

The *mixed* saliva is that usually employed for experimental purposes. The amount of this secretion in twenty-four hours is stated by BIDDER and SCHMIDT [1] to be 300 to 1500 c.c. The quantity secreted in the unit of time is largest during meals and falls to a minimum between meals. These variations will be discussed in greater detail later. The mixed saliva froths easily and is somewhat turbid, from admixture with epithelial cells and some leucocytes, together with the mucin obtained chiefly from the mucous glands of the mouth. The saliva contains vast numbers of bacteria.[2] The average specific gravity is 1.003 to 1.004. The reaction is ordinarily stated to be alkaline, but recent observations seem to indicate that it is neutral, perhaps even acid, in reaction. That saliva is able to neutralize strong acids does not indicate that this secretion is alkaline in reaction, but simply that it contains salts of strong bases combined with weak acids, such as carbonates and bicarbonates.

Toward litmus, lacmoid, and rosolic acid saliva reacts as though it were alkaline. When phenolphthalein is used, however, the solution remains colorless. Ten cubic centimeters of mixed saliva require 0.2 c.c. of a 1/10 normal alkali solution before the phenolphthalein will just turn pink, indicating an alkaline reaction. Saliva is therefore to be considered as at least neutral, if anything else, slightly acid, for phenolphthalein is in this case to be looked upon as a more reliable indicator than any of the others mentioned above. What holds for the

---

[1] VIERORDT: Tabellen, Jena, 1888, p. 128.
[2] See p. 160.

ordinary mixed human saliva is true also of the saliva obtained from the dog after stimulation of the chorda tympani nerve or after poisoning with curare.[1]

Among the inorganic constituents of the saliva are the chlorides, carbonates, bicarbonates, phosphates, and sulphates of sodium, potassium, magnesium, and calcium. The escape of carbon dioxide from the saliva on standing allows the precipitation of the bicarbonate of calcium as the carbonate. This substance is the chief constituent of salivary calculi and "tartar."

Much importance was at one time attached to the quantitative determination of the sulphocyanate in the saliva as an indicator of the rate of protein metabolism, for it is generally believed that this substance is a product of protein metabolism. MOORE, however, believes that its estimation is of little value.

The gases of the saliva are oxygen, nitrogen, and carbon dioxide. The latter predominates, and is found chiefly in chemical combination.

Among the organic constituents of the saliva the ferments *amylase* and *maltase* are of the greatest importance. The former of these two is characterized by its action upon starches which are under its influence converted into maltose. Maltase is especially effective in causing a still further cleavage of this sugar into dextrose. While food is being chewed, therefore, and for some time after it reaches the stomach, the starches contained in it are undergoing a change which converts them into substances readily absorbed by the alimentary tract. The mucin contained in the saliva is probably of little more than mechanical use in supplying food with a coating which allows the more ready passage into and through the œsophagus.

The œsophagus itself is studded with small mucous glands which secrete a small amount of a ropy fluid rich in mucin.

---

[1] See MUNK: Centralblatt für Physiologie, 1902, XVI, p. 33.

The function of this secretion is, so far as known, purely mechanical, in that it serves to lubricate the bolus of food as it passes through this tube.

2. **The Gastric Juice** is the term applied to the secretion of the mucosa of the stomach. Absolutely pure gastric juice can be obtained from human beings only in small quantities, impure juice in somewhat larger quantities. Pure juice has been obtained in cases of gastric fistula. Most observations are based upon a study of the impure juice obtained by feeding meals of known composition and then evacuating the contents through a stomach-tube or a fistula if such exists.

Pure gastric juice is best obtained from the dog by the method of "sham feeding" of PAWLOW and SCHUMOW-SIMANOWSKY.[1] This consists in feeding a dog in which the œsophagus has been separated in such a way from the stomach that the food never really enters the stomach. When such sham feeding is carried out, a reflex secretion of pure gastric juice occurs. Such a sham feeding may be kept up without injury to the dog for an hour daily, in the course of which time 200 to 300 c.c. of juice may be collected. The quantity secreted in twenty-four hours is subject to great variations, but amounts to about 1/15 to 1/10 the body weight in dogs. If we assume that the same proportion exists between the amount of the gastric juice and the body weight in the case cf the human being, a man weighing 70 kilos would secrete 4 to 7 liters daily.

KONOWALOFF[2] describes the gastric juice obtained from the dog by sham feeding as a clear, colorless, odorless liquid of a specific gravity averaging 1.00478. The acidity given by KONOWALOFF is higher than that of any other author, 0.544 percent of hydrochloric acid. A figure approximating this

---

[1] See Chapter XI, Part 1. PAWLOW and SCHUMOW-SIMANOWSKY: Centralbl. f. Physiol., 1889, III, p. 113. PAWLOW: Work of the Digestive Glands. Trans. by THOMPSON, London, 1903, p. 12.

[2] PAWLOW: l. c.

value was obtained long ago by HEIDENHAIN. In spite of the majority of observations to the contrary, this high acidity probably represents the true value under physiological conditions. Because of this acidity, bacteria do not develop well in gastric juice, and it may in consequence be kept for a long time without undergoing decomposition. The acidity of the *normal* gastric juice is due to *hydrochloric acid*. This was first proved by PROUT and independently of him, but later, by TIEDEMANN and GMELIN. The objections which were raised against this idea by CLAUDE BERNARD and BERRESWIL have been set aside by the researches of SCHMIDT.[1] Traces of organic acids are at times found in the normal gastric secretion.

Human gastric juice does not differ in its physical or chemical properties from canine juice. The concentration of the acid and the pepsin is perhaps somewhat lower in man.

No complete analysis of gastric juice seems to be at hand. Besides hydrochloric acid, gastric juice contains the ferment *acid-proteinase* (pepsin), which has the power of acting on proteins and splitting them into a number of simpler bodies. *Caseinase* (rennin) is also found in the gastric juice of the human being as well as that of other animals. This ferment has the power of curdling milk. The hydrochloric acid present in the stomach is of itself able to bring about this change, but the presence of the ferment can be demonstrated by neutralizing the gastric juice with an alkali, when it is found that the milk-curdling power still persists. It is lost, however, when the neutral gastric juice is heated to 70° C. for a short time. The secretion of the stomach also contains a fat-splitting ferment, *lipase* (steapsin), but its physiological importance is probably small, as the ferment acts little or not at all

---

[1] For an interesting account of the subject, see MOORE: Text-book of Physiology, edited by SCHAEFER, Edinburgh, 1898, I, p. 351.

in an acid medium of the concentration found in the stomach. A substance which is obtained as a coagulum from gastric juice on boiling probably represents the combined ferments, together with mucin and traces of organic material derived from particles of digested food or the dead cells of the stomach-wall itself.

The wall of the stomach contains the *antiproteinase* (antipepsin) discovered by WEINLAND. As its name indicates, this substance, when present, prevents the proteolytic ferments from acting. As it is probably due to the presence of this substance in the wall of the stomach, and in the intestines, that the alimentary tract does not digest itself, it will be discussed in some detail later.[1]

The inorganic constituents of the gastric juice exclusive of the hydrochloric acid consist of the chlorides and traces of the phosphates of sodium, potassium, magnesium, calcium, iron, and ammonium. The only figures indicative of the quantities in which these exist in the gastric juice are based upon analyses of the impure secretion or such as was obtained under other than physiological conditions. As these are practically valueless, they will not be given here.

The action of the ferments of the gastric juice, and the variations in its composition and quantity, will be discussed further on.

3. The Pancreatic Juice.—The quantity of juice which flows from the pancreas varies greatly in different animals and in the same animal under different physiological conditions. The observations of the older investigators are probably very incorrect. PAWLOW and his coworkers state that the amount of normal juice which can be obtained from a dog possessing a permanent fistula[2] is 21.8 c.c. per kilo in twenty-four hours. The pancreatic juice from a dog is a clear, colorless, odorless liquid, somewhat variable

---

[1] See p. 133.   [2] See Chapter XII, Part 1.

in composition and often containing enough protein to coagulate into a solid when heated.

The purest and apparently normal human pancreatic juice has been obtained by GLAESSNER [1] from a forty-six year old woman possessing a fistula of the pancreatic duct. GLAESSNER describes it as a water-like liquid which froths easily and on standing shows a very slight sediment. The juice has a decided alkaline reaction even toward phenolphthalein, so that it may really be regarded as alkaline in reaction. Analysis of two specimens showed respectively 1.2708 and 1.2494 parts of dry substance in 100 of the juice. Analysis of the dry substance showed, besides the ordinary inorganic salts, albumin, albumose, and peptone. GLAESSNER gives as the specific gravity of the juice 1.00748.

The quantity of juice secreted by GLAESSNER's patient in twenty-four hours varied from day to day, the extremes being 420 and 848 c.c. The quantity as well as the quality varied with the character of the food ingested. During starvation the quantity was very low, as was also the digestive power for proteins, fats, and starch. Under these conditions the alkalinity was also lower than after feeding.[2]

The pancreatic juice contains several ferments—*alkali-proteinase* (trypsin), *lipase* (steapsin), *amylase* (amylopsin), *caseinase* (rennin), and at certain times *lactase*. The *alkali-proteinase* is able to act upon proteins and to split them into a series of simpler substances in a manner similar to but more powerful than acid-proteinase. *Lipase* acts upon fats, breaking these up into fatty acid and alcohol. The *amylase* of the pancreas is similar to that of the saliva, and through its action on starch brings about the formation

---

[1] GLAESSNER. Zeitschrift für physiologische Chemie, 1904, XL, p. 465. For a review of the older literature on human pancreatic juice, see SCHUMM: Zeitschrift für physiologische Chemie, 1902, XXXVI, p. 292.

of maltose and dextrin. *Caseinase* is the name given to
the milk-curdling ferment of the pancreas. It is similar
to the milk-curdling ferment of the stomach. *Lactase* is
present in the pancreas only at certain periods in the life
of an animal and under certain conditions. It is found
normally in the pancreatic secretions of the puppy during
the period of lactation. In adult dogs it is absent, but it
can be made to appear again if the dog is kept for some
time on a milk diet or any diet containing milk-sugar. This
ferment has the power of acting upon milk-sugar (lactose)
and converting it into dextrose and galactose. A detailed
discussion of these ferments is given further on.[1]

It is as yet not entirely settled whether the ferments are
secreted as such from the pancreas or as proferments. According to DELEZENNE, FROUIN, and POPIELSKI,[2] the juice as
collected directly from the pancreatic duct never contains
alkali-proteinase (trypsin), but only its proferment (trypsinogen). This yields alkali-proteinase, however, as soon as
it flows over the mucous membrane of the duodenum, where
it encounters enterokinase.[3] Whether similar conditions hold
for the other ferments is a matter of dispute.

In human pancreatic juice GLAESSNER could find no alkali-
proteinase (trypsin), but only the proferment, which could,
however, be converted into alkali-proteinase by adding to
it an extract (enterokinase) of the mucous membrane of the
small intestine obtained from human corpses. Both lipase
and amylase were also found, but no maltase, sucrase, or
lactase, a fact which agrees well with experimental findings
in animals.

4. The Bile is the name given to the secretion from the
liver which is poured into the duodenum through the com-

---

[1] See Chapters IV, V, VI, VII, and VIII.

[2] DELEZENNE and FROUIN. Comptes rendus de l'acad., CXXXIV, p.
1526; Comptes rendus de Soc. biol., LIV, p. 691; POPIELSKI. Centralblatt für Physiologie, XVII, p. 65.

[3] See Chapter XIII, Part 4.

mon duct. The quantity of bile secreted in twenty-four hours varies not only in different animals but in the same animal under different circumstances. Contrary to the generally accepted view, the secretion of bile is not continuous and of varying intensity, that is to say, remittent, but intermittent. During certain periods no bile whatsoever escapes into the intestine. According to von Wittich and Westphalen, the total amount of bile secreted by a human being in twenty-four hours is about 500 c.c.[1] These observations were made on two men having biliary fistulæ. The specific gravity of the bile is, according to Westphalen, 1.0104. In 100 parts of liquid bile are contained 2.253 parts of solids.

All these figures must be looked upon as only approximately correct. Later,[2] when the quantitative and qualitative variations in the secretion of the bile are discussed, the reason for this will be more apparent. This explains also why the figures of scarcely any two of the score of observers who have worked on the biliary secretion agree.

The bile consists of the secretions of the liver-cells themselves, mixed with the mucus which is secreted by the bile-channels and the gall-bladder. As the bile passes from the liver-cells toward the intestine it undergoes a change, the bile from the gall-bladder and lower bile-passages being somewhat more viscid, because of loss of water and admixture with mucus, and cloudier, because of the presence of cells and cellular debris, than that collected from the bile-passages higher up. The color of the bile is different in different animals, and in the same animal, depending upon the conditions under which it has been obtained. Human bile obtained immediately after death has a golden-yellow, brownish-yellow, or at times slightly greenish color. That

---

[1] Vierordt: Tabellen, Jena, 1888, p. 135.
[2] See Chapter XIII, Part 1.

obtained at an ordinary post-mortem is yellowish brown, brown, or green. The reaction of the bile is ordinarily stated to be alkaline. It is probable, however, that it is neutral.

The most important constituents of the bile besides *water* are the *bile acids* in combination with the alkali metals, and the *bile pigments*. Among the other important constituents are the *inorganic salts* which are universally present in the body and *fats, lecithin, cholesterin, urea,* and *soaps*.

The bile acids are several in number and are usually divided into the *glycocholic* and *taurocholic* groups, each_of.which has several members. The bile pigments are also several in number. The reddish-yellow *bilirubin* and the green *biliverdin* are always present in bile under physiological conditions. *Hydrobilirubin* perhaps also belongs to this group. Under pathological conditions, as in gall-stones, a number of other pigments besides those already mentioned may be found, of which choletelin, bilifuscin, biliprosin, bilihumin, and bilicyanin are the most important. Bilirubin and hydrobilirubin are of great physiological interest because of the identity or at least close chemical relation of the former to a derivative of hæmoglobin, hæmatoidin, and of the latter to urobilin, one of the urinary pigments.

The inorganic constituents of the bile comprise the chlorides, phosphates, and sulphates of sodium, potassium, calcium, magnesium, iron, and copper. Urea is found only in traces. The following three analyses of human bile by HAMMARSTEN,[1] which probably are as trustworthy as any at our disposal, will give an idea of the relative proportions in which the various constituents exist in this secretion. The figures indicate parts per 1000.

Water.............	974.800	964.740	974.600
Solids.............	25.200	35.260	25.400

---

1 HAMMARSTEN: Text-book of Physiological Chemistry. Translated by MANDEL, New York, 1904, p. 276.

The latter consist of:

Mucin and pigments......	5.290	4.290	5.150
Taurocholate salts.......	3.034	2.079	2.180
Glycocholate salts........	6.276	16.161	6.860
Fatty acids from soaps....	1.230	1.360	1.010
Cholesterin............	0.630	1.600	1.500
Lecithin ⎫	0.220	⎧ 0.475	0.650
Fat ⎭		⎩ 0.956	0 610
Soluble mineral salts.....	8.070	6.760	7.250
Insoluble mineral salts....	0.250	0.490	0.210

The functions of the bile and the regulation of the biliary flow can be better discussed elsewhere.[1]

5. **The Intestinal Juice.**—Under this heading are grouped all the secretions of the intestine proper from the pylorus of the stomach to the anus. The intestinal juice has been obtained in man from cases of intestinal fistula; from animals it is obtained experimentally and unmixed with other digestive secretions or food by the production of either a THIRY or a VELLA fistula in any desired portion of the small or large intestine.[2]

The pure juice obtained from different portions of the intestinal canal below the stomach varies not only in amount but also in composition. We shall first consider the secretion of the small intestine (succus entericus). RÖHMANN [3] describes *the juice of the small intestine* of the dog as scanty in amount, viscid, and somewhat gelatinous in the upper portions, and as larger in amount and more fluid in the lower portions. THIRY [4] obtained a maximum secretion of 4 c.c. per hour from a loop of small intestine having a total surface of 30 sq. cm. The specific gravity is given by the same author as 1.0115, and the juice is described as clear, light yellow, somewhat opalescent, and of a decided alkaline reaction. The juice from the small intestine contains

---

[1] See Chapter XIII, Parts 1 and 2.
[2] See Chapter XIII, Part 3.
[3] RÖHMANN. Pflüger's Archiv, 1887, XLI, p. 424.
[4] THIRY. Vierordt's Tabellen, Jena, 1888, p. 140.

protein, mucin, urea, and the ordinary salts found everywhere in the secretions and tissues of the body. The amount of carbonate is particularly high, to which fact the alkaline reaction of the juice is attributed.

The descriptions given of the secretion of the small intestine in the human being are practically the same as those given above for the juice obtained from dogs. The best observations on human intestinal juice are those of TUBBY and MANNING,[1] HAMBURGER and HEKMA,[2] and NAGANO.[3] In TUBBY and MANNING's case pure intestinal juice was obtained from a piece of intestine 3½ inches long situated some 8 inches above the ileocæcal valve. Their description of the juice. is similar to that of other observers and may be taken as a type. The daily yield of juice from this piece of intestine varied from 19 to 35 c.c.; the specific gravity from 1.0016 to 1.0162, on the average 1.0069. The fluid was opalescent, and on standing a sediment consisting of leucocytes and desquamated intestinal epithelium formed. The juice gave a strong alkaline reaction, and contained protein, mucin, and the ordinary salts, of which the carbonates and chlorides of sodium and potassium were the more conspicuous.

Within the last few years our conceptions of the nature of the chemical substances contained in the secretions of the intestine and in the walls of this portion of the alimentary tract have undergone great revision, and from having looked upon this chapter of alimentation as comparatively unimportant we now rank it with chapters on the pancreas and stomach. This is due to the discovery in the intestinal juice and in the walls of the intestine of a number of new chemical substances, an understanding of the functions of which, together with a proper appreciation of the physio-

---

· TUBBY and MANNING: Guy's Hospital Reports, London, 1891, XLVIII, p. 277.

² HAMBURGER and HEKMA: Journal de Physiol., IV. p 805.

³ NAGANO: Mittheilungen aus den Grenzgebieten der Medicin und Chirurgie. IX, p. 393.

logical importance of those already well known, has given us
a new insight into the importance of the small intestine in
the great problem of alimentation. The following is a list
of these chemical substances, together with a brief indication
of their physiological functions, which are discussed in greater
detail later.

Proteinase. This ferment is found in the secretion and
mucous membrane of the duodenum and originates from
the glands of BRUNNER, found in this section of the intestine.
It seems to be identical with the acid-proteinase of the
stomach, as it acts best in an acid medium. It is evident
that some opportunity is given this ferment to aid in the
digestion of proteins. But as compared with the activity of
the stomach, or pancreas, that of the duodenum is probably
only small.

Protease (erepsin) is one of the most important ferments
found in the intestinal juice. This ferment, discovered by
COHNHEIM, has no action upon "native" proteins except
casein, but has the power of splitting proteoses and peptones into a number of simpler substances (mono- and
diamino acids, ammonia, etc.) which are similar to those
formed through the action of either acid- or alkali-proteinase on proteins. Protease must therefore not be con-
founded with the proteolytic ferments of either the stomach
or pancreas. The presence of this ferment in human intestinal juice has been proved by HAMBURGER and HEKMA.

Lipase (steapsin). It is ordinarily stated that lipase does
not occur in the intestinal juice or in the mucosa. The
experiments of KASTLE and LOEVENHART indicate, however,
that it is found in all portions of the mucosa of the small
intestine in sufficient amounts to be of the greatest importance in the absorption of fats. The lipase of the intestine
is identical with that of the pancreas and pancreatic juice,
and, like the latter, splits fats into fatty acid and alcohol.

Amylase is found in the small intestine and is secreted
in sufficient amounts to play an important rôle in the absorp-

tion of starches. The ferment splits starches into malt-ose and dextrin. Its presence in human intestinal juice has been proved by NAGANO, HAMBURGER and HEKMA.

Maltase. The occurrence of this enzyme in the small intestine is still questioned. Evidence is slowly accumu-lating, however, to show that it is to the presence of this enzyme in the intestinal juice and the intestinal mucosa that we owe the splitting of the maltose formed from starch into the dextrose found in the intestine after the ingestion of starch or malt-sugar. The presence of a maltose-splitting ferment in human intestinal juice is indicated by the re-searches of NAGANO, HAMBURGER and HEKMA.

Sucrase (invertase, invertin) is one of the most important enzymes found in the secretion of the small intestine. Cane-sugar, which serves as such a common article of diet, would without the presence of this enzyme be scarcely absorbed by the intestine. The sucrase splits the cane-sugar into the readily absorbable dextrose and lævulose.

Lactase is found in the intestinal tract of infants and those adults who consume milk-sugar either as a separate article of diet or as one of the constituents of milk. Lactase acts upon milk-sugar (lactose) and splits this into dextrose and galactose.

Arginase is a ferment which has recently been discovered by KOSSEL and DAKIN. It has the interesting property of splitting arginin into the two chemically much simpler sub-stances ornithin and urea. It is a ferment which occurs not only in the mucous membrane of the intestine, but also in a number of the parenchymatous organs.

Antiproteinase (antipepsin and antitrypsin). This sub-stance is found in the mucosa of the intestine, but not in the secretions of the intestine. Like the antiproteinase of the stomach, it has the power of preventing the proteinases from acting upon proteins and splitting them into their well-known decomposition-products.

Enterokinase. The ferment character of this substance

has not as yet been entirely established. It is the name given by PAWLOW to a substance, discovered by CHEPO-WALNIKOW in the mucous membrane and secretions of the small intestine, which has the power of converting the inactive proferment of the pancreas into the active alkali-proteinase (trypsin). HAMBURGER and HEKMA have found this substance in human intestinal juice.

Pancreatic Secretin is in no sense a ferment. It is a term applied by BAYLISS and STARLING to a substance formed in the upper portion of the intestine during digestion, which is absorbed into the blood and, reaching the pancreas, increases the secretion from this gland.[1]

The secretion of the large intestine has been studied in the human being in cases of fistula opening into the ascending, transverse, or descending portions of this part of the intestinal tract, and in animals in which artificial fistulæ have been created. The observations of all who have worked on the secretion of the large intestine agree in stating that it is small in amount and mainly composed of mucus. The juice is alkaline in reaction and seems to contain no enzymes of digestive importance. It is probable, however, that the food which escapes through the ileocæcal valve, mixed with the enzymes poured out upon it higher up in the alimentary canal, continues to undergo digestive change in the large bowel. Of greatest importance is the absorption which occurs in the various portions of the colon. While the food is soft in consistency in the ascending portion of the large intestine, it becomes less liquid and finally firm as the transverse and descending portions are reached, until in the rectum the solid fæces are formed.

While no enzymes derived from the large intestine itself act upon the food in the colon, enzymes derived from the bacteria present here bring about most profound changes in the food.

---

[1] The ferments are considered in greater detail in the succeeding chapters. Enterokinase is discussed in Chapter XIII, Part 4, pancreatic secretin in Chapter XII, Part 5.

These are chiefly of a putrefactive character, and affect in the main the protein constituents of the food. Certain of the bacterial enzymes belong to the proteinase group, and these split the undigested protein remnants into leucin, tyrosin, and the other products of proteinase digestion. Indol, skatol, and various phenols are produced also, as well as fatty acids (lactic, butyric, caproic, etc.) and gases (hydrogen, hydrogen sulphide, and methane).

CHAPTER IV.

FERMENTS AND FERMENTATION.

1. Organic Ferments.—Since ferments, the activities of which we recognize in so many physiological processes, play a great rôle in that special chapter of physiology with which these pages deal, namely, alimentation, it may not be amiss to discuss here their general properties.

Ferments represent only a special class of the so-called *catalytic agents*, and are in consequence distinguished by the same characteristics as catalyzers in general. *A catalyzer is any substance which, without appearing in the end-products of a chemical reaction, alters by its mere presence the rate of this chemical reaction.* The catalytic agent is therefore not to be looked upon as the *cause* of the chemical reaction, for this occurs even in the absence of the catalyzer, only then at a different rate.[1] When a catalytic agent *increases* the velocity of a chemical reaction it is said to be a *positive catalyzer;* on the other hand, when it *decreases* the rate of a chemical reaction it is known as a *negative catalyzer.* Examples of positive catalyzers will occur to every one. Common ones from *inorganic* chemistry are manganese dioxide, which hastens the decomposition of potassium chlorate into potassium chloride and oxygen when heated; and nitrogen tetroxide, which accelerates the oxidation of sulphurous acid into sulphuric in the manufacture of the latter substance. In both instances

---

[1] OSTWALD: Über Katalyse, Leipzig, 1902, p. 12. COHEN: Physical Chemistry for Physicians. Translated by MARTIN H. FISCHER, New York, 1903, p. 35.

the catalyzers do not appear in the end-products, and may, when the reaction has been brought to a standstill, be recovered in an unaltered state. As is well known, the decomposition of potassium chlorate and the oxidation of sulphurous acid occur even when the catalytic agents are not present, but much more slowly.

Any of the ferments may be cited as examples of *organic* catalyzers. We may mention here lipase, which hastens the chemical decomposition of fat into fatty acid and alcohol, and proteinase, which accelerates the rate of decomposition of proteins into simpler substances. Here also we deal with chemical reactions which take place even when no ferments are present. Under such circumstances the decompositions occur very slowly, however, requiring weeks, months, or years to attain the degree of decomposition which in the presence of the respective ferments may be reached in a few hours. As was found to be the case with inorganic catalyzers, so here also we find that the ferments do not appear in the end-products of the reactions which they catalyze.

Examples of *negative* catalyzers are much more difficult to discover. Only isolated ones have been described, and since nearly all of them seem now to be regarded as inhibitors of positive catalyzers, they will not be discussed here. Negative *organic* catalyzers are entirely unknown, unless we look upon the *antiferments* as belonging to this group. Whether they do or not must be left undecided until the mode of action of the antiferments has been discovered. These substances decrease markedly the velocity of certain chemical reactions *occurring in the presence of a ferment*. Until we know whether the antiferment produces its effects by combining with the ferment (either chemically or mechanically) or by decreasing directly the velocity of the chemical reaction, the classification cannot be made. Only in case the latter of these two possibilities were proved to be the correct one could we look upon the antiferments as negative catalyzers.

The catalytic agents produced in living cells or tissues are called *ferments*. They are usually divided into two groups; the so-called *organized* and *unorganized* ferments. The first term is applied to those ferments which are connected in some way with the life of the cells in which they are produced, and which cannot be extracted from these cells. The unorganized ferments can, on the other hand, be extracted from the cells in which they are formed, and are able to produce their characteristic actions outside of the cells as well. The unorganized ferments are also known as *enzymes* (KÜHNE). To the latter class belong all the digestive ferments, amylase, maltase, acid- and alkali-proteinase, etc., which it has been possible to extract from the tissues or secretions of the alimentary tract, and which produce their characteristic reactions in a test-tube as well as in the living intestinal tract. To the class of organized ferments belong all those whose presence we recognize only by their chemical behavior and whose activities continue only as long as the cell in which they were produced is intact. Many of the "vital" activities of tissues have to be attributed to such organized ferments. It is useless to enumerate the organized ferments, because the list of unorganized ferments (enzymes) is constantly growing at the expense of the organized, and we may hope in time to speak of enzymes alone. To illustrate, we need only cite BUCHNER's successful extraction of zymase from the yeast-cell, which is able to bring about the decomposition of glucose into alcohol and carbon dioxide quite as readily as the yeast-cell itself. The discovery of BUCHNER is mentioned in this connection because it takes away one of the pillars which for decades have been utilized to support the idea of the essential difference between organized and unorganized ferments. In the pages which follow the terms ferment and enzyme are used synonymously.

The number of known ferments is already very large, and is being added to daily. It is unfortunate that a uniform system of nomenclature has not yet been adopted by all the

workers in this field.   Persistence in the use of old methods
of naming even newly discovered enzymes, and the existence
of the same ferment under different names, are extremely
confusing.   The best method of naming ferments to-day
is to add the ending *ase* to the root of the Latin or Greek name
of the substance upon which the ferment acts.   Thus maltase
acts on maltose, sucrase on sucrose, tyrosinase on tyrosin,
amylase on starch, etc.   Since the action of many, probably
all, ferments is *reversible* (as will be explained later), and as
in this way several names might be given to one and the
same ferment, it should be the rule to add the ending *ase*
to the root of the chemically more complex substance upon
which the ferment acts.   The ferment which accelerates the
decomposition of sucrose into "invert-sugar" is therefore
better named sucrase than invertase, for sucrose is chemically more complex than the dextrose or the lævulose which
constitute the invert-sugar.   It is sometimes convenient to
add the ending *ase* to the Latin or Greek root of a word express-
ive of some striking characteristic of a group of ferments.
Under the *oxidases*, for instance, are classed a large number
of individual enzymes, all of which have the power of catalyzing the oxidation of various chemical substances.   But there
are many specific oxidases, and these are regrouped under
the general heading, according to the substance or substances with the oxidation of which they are particularly
concerned, as tyrosinase, laccase, olease, etc.

Sometimes ferments having the same specific activity
show differences in their behavior toward altered external
conditions.   For example, the lipase (steapsin) obtained
from the pancreas does not behave in exactly the same way
toward changes in temperature and concentration, toward
acids, alkalies, etc., as the lipase obtained from certain vegetable cells.   This has aroused the suspicion that they may not
all be identical.   For this reason ferments are often spoken
of in the plural, as lipases, meaning thereby fat-splitting
enzymes derived from any source.

We are only just beginning to get some idea of the chemical composition of a few of the simpler ferments. For this reason the recognition of the existence of a ferment within a cell or out of it is still dependent, not upon the recognition of the chemical substance itself, but rather upon the discovery of certain properties common to all ferments and specific ones characteristic of separate enzymes. The following are usually looked upon as properties common to all ferments, and the first four are utilized in proving the existence of a ferment in cells or tissues, or in extracts made of these.

(*a*) *A ferment does not initiate a chemical reaction but only alters its velocity.* The ferment acts essentially through *contact*, in that its mere presence is responsible for the altered velocity of the catalyzed reaction. At the end of the reaction the ferment is (under ordinary circumstances) found in an unaltered state in the reaction mixture. The ferment does not, therefore, enter into the end-products of the reaction. When hydrochloric acid is poured upon sodium carbonate in order to decompose it, we have at the end of the reaction the chlorine of the hydrochloric acid appearing in the sodium chloride, and the hydrogen in the carbonic acid. Not so, however, in the case of a ferment. When sucrase (invertase, invertin) acts upon cane-sugar, dextrose and lævulose are formed, but neither of these products contains in chemical combination any part or all of the sucrase molecule,—the sucrase is found in an unaltered condition at the end of the reaction.

It is possible that the molecules of the ferment enter *temporarily* into chemical combination with the substance acted upon or with its products. This assumption is made on the ground that in some simple catalytic processes the catalyzer does temporarily combine with the reacting substances. This is the case, for example, in the manufacture of sulphuric acid, in which steam, sulphur dioxide, oxygen, and nitrogen dioxide are introduced simultaneously into a large chamber. In brief, it is believed that the sulphur dioxide and steam combine to form sulphurous acid, which is oxidized to sulphuric by the nitrogen tetroxide formed through the union of the nitrogen dioxide with oxygen. The nitrogen dioxide, which plays the rôle of a catalyzer in this reaction, unites first of all with oxygen to form nitrogen tetroxide, but the nitrogen dioxide is restored immediately in that the sulphurous acid takes away the oxygen from the nitrogen tetroxide. The nitrogen dioxide acts, therefore, only through its mere presence, appearing in an unaltered state at the end of the reaction. Nor does it initiate a chemical reaction, for the oxidation to sulphuric acid occurs whenever sulphurous acid is exposed to oxygen. But the reaction, which under these circumstances occurs only very slowly, occurs very rapidly when nitrogen dioxide is present.

There are other facts, into a discussion of which we cannot go here, that speak against this conception of the action of a · ferment, and seem to indicate that certain physical characteristics of the ferment (or, in general, any catalytic agent), such as its surface, electrical condition, etc., constitute the real cause of its peculiar action. However this problem may ultimately be settled, the fact seems fairly well established that, except for a simple decomposition which is to be discussed later, the ferment appears in an unaltered state at the end of the reaction.

(b) *In infinite time the amount of chemical change brought about by a ferment is independent of the concentration of the ferment.* This means that a small amount of a ferment will bring about as much chemical change as a larger one, provided unlimited time is given. In this regard, therefore, a ferment differs from the ordinary constituent of a reaction mixture. ▪ We know from the law of chemical mass-action of GULDBERG and WAAGE that in any ordinary chemical reaction the amount of the chemical change is proportional to the concentration of the reacting substances. This does not hold in the case of ferments, where the amount of ferment is in nearly all cases exceedingly small when compared with the amount of substance acted upon. This fact alone speaks against any idea that fermentation is dependent upon a quantitative chemical decomposition between catalyzer and catalyzed substance—that, in other words, we are dealing with a stoichiometrical reaction.

It is well to note that what has been said holds true only when infinite time is allowed. *For shorter periods*—which vary with the nature of the catalyzer and the nature of the substance or substances catalyzed—*the degree of fermentation is clearly and often definitely a function of the concentration of the ferment.* We cannot here enter into a discussion of the influence of the concentration of a ferment upon the velocity of the fermentation. In general it can be said that only within exceedingly narrow limits of time and concentration, both of the substances undergoing catalysis and the catalyzer, is the rate of fermentation approximately directly proportional to the concentration of the ferment. In the majority of instances the variation is very great, and an increase in the concentration of the ferment is not followed by a corresponding increase in the rate of the fermentation. Thus doubling the concentration of a ferment in a certain reaction mixture does not multiply the rate of the catalysis by two, but by less than two.

The reason why the rate of catalysis is not proportional to the amount of ferment added is not yet entirely clear. It seems to be dependent, in part at least, upon a decomposition which the ferment itself suffers, in consequence of which its concentration is steadily decreased. More potent than this seems to be the accumulation of the products of the fermentation. The opinion has been expressed that the accumulation of the products of the catalyzed reaction injures the ferment, but this view can no longer be held, since we have learned that the activity of many enzymes is reversible. For example, in the action of maltase on maltose and the production of dextrose we might assume that the ever-increasing amount of glucose in the reaction mixture interferes with the further action of the maltase. The argument falls to the ground, however, when we deal with the action of maltase on glucose and the production of maltose. If glucose interfered with the action of the enzyme, the velocity of this reaction ought to be lowest at first and increase as more and more maltose is produced. As an actual matter of fact, the opposite occurs, and the velocity of the reaction becomes progressively less as the maltose accumulates. In either case, therefore, the accumulation of the fermentation-products seems to interfere with the chemical reaction itself.

(c) *Ferments are usually characterized by a great sensitiveness to comparatively low temperatures.* In fact, it is one of the commonest means employed in proving that a certain chemical reaction is dependent upon the presence of a ferment, to heat the reaction mixture to boiling and find that after this treatment the reaction no longer takes place. (In most instances the reaction does in reality still occur, only so infinitely slowly, as a general thing, that it is unrecognizable by the analytical methods at our disposal.) This means that a temperature of 100° C. destroys the activity of the ferment. Some few ferments are able to endure this temperature or even a higher one for a short time, but the vast majority cannot stand heating to even 60° C. for any length of time. The cause of this inactivation of the ferment is to be sought in a decomposition which it undergoes. The nature of this decomposition is not as yet understood. Ferments are not nearly so sensitive to temperature in the dry state as in the moist. Many ferments which are readily destroyed by heating to 60° C. for some minutes in the presence of water withstand a temperature of 120° C. for much longer periods if the ferments are thoroughly dry. An every-day illustration of this is found in the well-known fact that in surgical sterilization the death of all bacteria and spores is obtained much more easily by the use of moist heat than dry heat. These facts seem to indicate that the decomposition of the ferment is accomplished through the taking up of water—in other words, that we are dealing with a hydrolysis of the ferment. The possibility of a mere change in the physical state of the ferment as the cause of its inactivation through heat must, however, also be considered, for there are many reasons at hand for believing that ferments owe their specific virtues quite as much to their physical state of aggregation as to their chemical make-up.

The fact that ferments suffer a decomposition when exposed to higher temperatures explains the peculiar behavior of chemical reactions which are being catalyzed by ferments when the reaction mixtures are exposed to various temperatures. As is well known, the velocity of ordinary chemical reactions varies greatly with different temperatures. An every-day illustration of this fact is found in the use of heat to hasten chemical decomposition in our ordinary analytical reactions, and the resort to refrigeration in the effort to delay decomposition in animal and vegetable matter. For ordinary chemical reactions it has been found that the reaction velocity is, roughly speaking, doubled or trebled for every increase of 10° C. in temperature.[1] When we are dealing with chemical reactions which are taking place under the influence of a ferment, things are entirely different. Within certain limits we have here also an increase in reaction velocity with an increase in temperature, but when a certain point is reached—which varies with different ferments, and with the same ferment acting in different media—the reaction velocity no longer increases but *decreases* with a further elevation of the temperature. Finally a point is reached at which the reaction ceases entirely (apparently). It is for this

---

[1] COHEN: Physical Chemistry for Physicians. Translated by MARTIN H. FISCHER. New York, 1903, p. 53.

reason that an *optimum temperature* exists in the case of all ferments. By this is meant the temperature at which the particular ferment under the given conditions brings about its characteristic effects most rapidly—that is to say, its reaction velocity is greatest. This optimum temperature lies, for most ferments, close to 40° C.

The reason for the existence of such a maximal reaction velocity can be readily understood from what has been said before. If the ferment remained unchanged upon heating a reaction mixture, the velocity of the chemical reaction would increase progressively with an increase in temperature, just as in any ordinary reaction mixture. When, however, we are dealing with reactions in which ferments are concerned, the ferment is all the time undergoing a decomposition itself, and the rate of this decomposition is also a function of the temperature. It is known, in general, that the length of time required to destroy a ferment at a comparatively low temperature is greatly reduced by an elevation of only a few degrees centigrade. TAMMANN, who has worked out the velocity of the decomposition of the ferment synaptase (emulsin) above 60° C., finds that an increase in temperature of 10° C. multiplies the decomposition velocity of the ferment by more than seven. In a reaction mixture in which a ferment is concerned we have therefore at least two reactions going on side by side, and it is the product of these two simultaneously occurring reactions which is represented by a curve that attains a maximum at one point. Since the two reactions which yield this curve have their individual characteristics, it can be readily understood why the curves not only of different enzymes, but even of the same enzyme acting under different external conditions, must vary. In Fig. 18 are shown the curves representing graphically the changing reaction veloc. ities of various ferments with variations in the temperature. Curves 1, 2, 3, and 4 illustrate the behavior of indigo enzyme obtained from various sources; curve 5 that of synaptase (emulsin) obtained from sweet almonds. The curves show very well a region in which an increase in temperature is followed by an increase in the rate of chemical reaction, the attainment of an optimum, and the rapid fall to the base-line, with only a slight further increase in the temperature beyond the optimum. In general, it may be said that all ferments behave in a similar manner.[1]

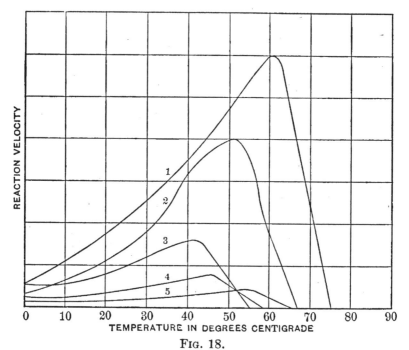

Fig. 18.

(Copied from COHEN: Physical Chemistry for Physicians. Trans. by FISCHER, New York, 1903, p. 56.)

(d) *The reactions which are catalyzed by a ferment are rarely, if ever, complete unless the products of the reaction are removed as soon as formed.* By this is meant that if a certain substance and a ferment capable of acting upon this substance are mixed together, say in a test-tube, not all of the substance will be acted upon by the ferment, but when the

---

[1] COHEN: Physical Chemistry for Physicians. Translated by MARTIN H. FISCHER, New York, 1903, p. 55.

reaction has come to a standstill a certain amount of **the** substance will be found unchanged in the reaction mixture. If, for example, a certain amount of fat and lipase (steapsin) are mixed together and the whole is allowed to stand until the reaction has come to a standstill, it is found that some of the fat has been left in an undigested state. What has been said of fat holds true also for a large number of other ferments—in fact, all that have thus far been investigated.

It is only recently that an explanation of this interesting and fundamental fact has been given, and, as will become apparent later, a large number of physiological processes rendered more intelligible in consequence. It used to be said that the products of the fermentation interfered with its further progress. We know now, however, that the reason for the incompleteness of the chemical reaction lies in the fact that the ferment *is continually re-forming from the products of the reaction the substance or substances with which the ferment was mixed originally.* Thus, in the case of fat, which is split into fatty acid and alcohol by lipase, we have two reactions going on side by side: first, the long-recognized *analytical* one, by which fatty acid and alcohol are formed from the fat; and, second, a *synthetical* one, by which fat is formed from the fatty acid and alcohol. Both reactions are catalyzed by the same ferment, the action of which we say is reversible. *This reversibility of the action of a ferment is another of its fundamental characteristics.* Whether in any given case a ferment acts synthetically or analytically, it will be shown later, is determined solely by the ordinary laws of chemical equilibrium.

If the products of a chemical reaction which is being catalyzed by a ferment are removed as soon as formed, the reaction will be completed. If, in the illustration cited above, the fat and lipase are put, not into a test-tube, but into a parchment bag suspended in a current of water, all the fat will be split into fatty acid and alcohol. Under these circumstances the fatty acid and alcohol diffuse through the parchment wall as soon as formed and are carried off by the stream of water: The products cannot therefore accumulate and be synthesized into fat. In this way all the fat is ultimately split into fatty acid and alcohol.

(e) For reasons which are at present not well understood, *the fermentative activity of extracts of organs or the isolated ferments of the alimentary secretions is disappointingly low when compared with the activity of the uninjured organ or the freshly obtained secretion.* The fact that all the ferments thus far studied are apparently colloidal in character, and consequently exceedingly sensitive to changes in their surroundings, is perhaps responsible for this, at least in part. Some very active preparations of ferments have, however, been described. The most active are, no doubt, the preparations of amylase prepared by WEINLAND.[1] The following two experiments, taken from his note-books, may serve as illustrations.

(α) Forty-one grams of the mixed pancreas from three pigs are ground up for ten minutes in a mortar with 41 grams of quartz sand. To the mixture are added 21 grams of infusorial earth, and the whole is ground for another ten minutes. A paste results. This paste is placed in a cloth and subjected to the pressure of a BUCHNER press. Only 7 c.c. of juice (1) are obtained. On the following day the paste is mixed with 30 c.c. of 0.9 percent NaCl solution and 10 c.c. of disodium phosphate solution. This gives a somewhat thin mixture. This is again put under the press and 29 c.c. of juice are obtained, to which are added the 7 c.c. (1) obtained before. A third pressing raises the entire amount of extract to 43 c.c. (in other words, about the volume of the original amount of pancreas). The extract is filtered, when a clear, yellowish solution results, and this is covered with toluol to prevent the development of bacteria. Five days later 5 c.c. of a two-percent glycogen solution (upon which amylase acts as readily as upon starch) are mixed at room temperature with a little KI-I solution, when the whole assumes a dark-red color, indicating the presence of the glycogen. To this mixture there

---

[1] WEINLAND: Personally communicated.

is added 0.1 c.c. of the extract prepared above. *In one minute the glycogen solution has lost its color,* and the TROMMER test results positively. The paste from which the extract has been obtained fails to remove the color from a similarly prepared mixture of glycogen and iodine, even when several minutes are allowed.

(β) By methods similar to those described above, 60 c.c. of juice are obtained from 59.4 grams of dogs' pancreas. The extract has been kept for five days under toluol. When 2 c.c. of this extract are mixed at room temperature with 1 c.c. of a two-percent glycogen solution, and to this is added *as quickly as possible a KI-1 solution, no change in color is obtained,* indicating that the glycogen has been changed into sugar even in these few seconds. Even when 3 c.c. of the glycogen solution are added to the extract, the decomposition of the glycogen occurs so quickly that a reaction with iodine cannot be obtained. *Not until 3 c.c. of the extract are mixed with 25 c.c. of the glycogen solution (½ gram dry glycogen) is this possible.* After this mixture has been kept in the incubator at 37° C. for one-half hour, it, too, no longer gives a positive iodine reaction. On the following day 0.5 c.c. of the extract is mixed with 5 c.c. of the glycogen solution. After standing for eight minutes at room temperature the mixture has lost its color. The TROMMER test is positive. Examination of the paste from which the extract has been obtained shows it to be unable to act upon the glycogen.

(*f*) It seems to be true of a number of the ferments found in cells and in their secretions that these ferments do not exist as such in them, but in an inactive form known as *proferments* or *zymogens.* The most striking example of such a zymogen is probably the entirely inactive proferment of trypsin (trypsinogen, as it is sometimes called), found in the pancreatic juice. This proferment (as present in the pancreatic juice obtained directly from the pancreatic duct) is not able to act upon proteins until it has been converted into trypsin (alkali-proteinase) by coming in contact with a substance contained in the secretions of the small intestine (enterokinase). The proferment of alkali-proteinase is found also in the body of the pancreas, and this fact has been considered in trying to explain why the pancreas is not digested by its own trypsin, or in general why any organ is not digested by the ferments contained in it. This explanation of the immunity of tissues against their own ferments can hold in only a few cases, however. For other enzymes the existence of proferments has not been so definitely established.

There seems to be little doubt but that proferments exist for rennin (caseinase), steapsin (lipase), and pepsin (acid-proteinase). The existence of a proferment of acid-proteinase in the gastric mucosa is demonstrated by the following facts: If the minced mucous membrane of the stomach of an animal is extracted with a dilute alkaline solution and the whole is filtered, a clear solution is obtained which does not act on proteins. If this solution is acidulated with hydrochloric acid it digests proteins rapidly. But if this acid solution is rendered alkaline once more, and is then again acidulated with hydrochloric acid, the digestive properties of the solution for proteins will have been lost. These experimental facts are explained on the assumption that pepsin is readily destroyed in an alkaline medium, while its zymogen is not, and that the conversion of the zymogen into the ferment is accomplished through the acid.

The relation that ferments bear to their proferments is not-understood. The conversion of the proferment into the ferment may at times be accomplished through the action of a specific substance, such as enterokinase. But the character of this substance, which by some is regarded as itself a ferment (whence the name), by others as a substance differing markedly from this group of compounds, is also not understood. In the majority of instances dilute acids are able to bring about a conversion of the proferment into the ferment. Whether this represents more than a simple change in the physical state of a colloid is still unknown.

2. Inorganic Ferments.—Beyond what has already been said we cannot here enter into a discussion of the various theories of fermentation which have been proposed from time to time. Within the last few years, however, the work which has been done on the so-called *inorganic* ferments has so altered our older conceptions of the essential nature of fermentation that a discussion of some of the analogies between the organic and inorganic ferments may not be amiss, even though the facts as they stand now have a greater bearing on the theoretical than on the practical side of the medical problems of fermentation.

As early as 1863 SCHÖNBEIN pointed out the analogy which exists between the influence of finely divided metals (such as platinum, iridium, silver) and that of a number of ferments on the velocity of certain chemical reactions. The decomposition of hydrogen peroxide, for instance, is accelerated not only by the addition of aqueous extracts of various animal and vegetable cells which contain ferments, but also through the addition of the finely divided noble metals mentioned above. The decomposition of hydrogen peroxide occurs also in the absence of these substances, only then much more slowly, so that the finely divided metals possess the characteristic property of a ferment, namely, that of altering the velocity of a reaction which occurs in its absence also.

Through the more recent work of BREDIG and with him VON BERNECK, REINDERS, IKEDA, and the efforts of McINTOSH, BILLITZER, NEILSON, and others, our knowledge of the inorganic ferments has been greatly increased. For a fuller account than can be given here and for references to the literature the reader is referred to the monograph of BREDIG.[1]

In place of the coarsely powdered metals employed by the older observers, BREDIG used so-called *colloidal* or *pseudosolutions* of these metals. The solutions are prepared by sending an electric current through electrodes of the noble

---

[1] BREDIG: Anorganische Fermente. Leipzig. 1901; Ergebnisse d. Physiologie, 1902, I, 1te Abth., p. 134.

metals while these are held in a large dish of absolutely pure water. The metal emanates in a cloud from one of the electrodes and remains *suspended* in the liquid. The colloidal solution is therefore not a true solution, but rather a suspension of very fine particles. The coarser particles are filtered off, and the *sol* (a term applied to a colloid in the liquid state) which remains behind shows properties exceedingly like those of the ordinary ferments. For this reason these sols have been called *inorganic ferments* by BREDIG. Depending upon the metal from which the sol is prepared, we speak of platinumsol, goldsol, etc.

A large number of ferments from different sources have the power of catalyzing the decomposition of hydrogen peroxide into water and oxygen. Colloidal solutions of platinum have the same power. Interestingly enough, as is the case with the true ferments, this inorganic ferment also is active in exceedingly small quantities. Thus one gram-atom of platinum (194.8 grams) diluted with seventy million liters of water is still able to accelerate the decomposition of more than a million times its amount of hydrogen peroxide. One cubic centimeter of the solution which still shows "fermentative" properties therefore contains only 1/300000 milligram of platinum. It would be difficult to rival this with figures taken from any of the organic ferments.

The rapidity of the decomposition of hydrogen peroxide by platinumsol is dependent upon the concentration of the platinum, just as in the case of a true ferment. If infinite time is allowed, a small amount of platinumsol will bring about as much decomposition as a larger amount. In shorter periods the velocity of the catalysis is definitely dependent upon the concentration of the colloidal platinum, very much as in the case of the ordinary ferments which have the power of catalyzing the decomposition of hydrogen peroxide.

Even though not sensitive to the same degree as organic ferments, the inorganic ferments are exceedingly sensitive to heat. In the preparation of the sols great care has to be exercised to prevent overheating of the water, as this causes the platinum, gold, silver, or whatever metal is being employed, to be precipitated. In other words, the inorganic ferment is destroyed. For this reason, therefore, the dish of water in which the colloidal solution is being prepared is kept cooled with ice, and overheating by allowing the electric current to pass through the water for too long a time is carefully avoided. It was pointed out above that the decomposition of the true ferments by heat might represent nothing but a physical change. For, as far as we know now, the organic ferments are all colloidal solutions, and in consequence exceedingly liable to precipitation (and inactivation) through heat. The greater sensitiveness to heat in the case of the true ferments is readily explained on the ground that these are always impure—mixed with salts, etc., from which it is impossible to free them—and consequently more readily precipitated. As will be shown in greater detail further on, the presence of certain impurities, such as gases and salts, in the colloidal solutions of platinum greatly decreases their stability also. Colloidal solutions of platinum, gold, etc., are the more readily precipitated (destroyed) by heat the greater the amount of these impurities. It may be for this reason that the purest ferments which have thus far been obtained are least sensitive to temperature. Whether inorganic ferments have an optimal temperature as do the true ferments is not yet worked out sufficiently. The recent work of ERNST [1] shows that such a temperature exists in the case of the decomposition of oxyhydrogen gas in aqueous solution in the presence of colloidal platinum.

It was held for a long time that the inorganic ferments differed from the organic in one essential detail. While the action of the ordinary ferment comes to a standstill after a certain time, a reaction catalyzed by an inorganic ferment seemed always to be complete. It was shown above that

---

[1] ERNST: Zeitschr. f. physik. Chem., 1901, XXXVII, p. 448.

the explanation of this partial decomposition by organic ferments lay in the fact that their action was reversible, that they synthesized from the products of the decomposition the substance which was being analyzed. Now it has been shown by Neilson [1] that finely divided platinum is able not only to hasten the splitting of ethyl butyrate into ethyl alcohol and butyric acid, but is also able to synthesize ethyl butyrate from ethyl alcohol and butyric acid. *Reversibility* is therefore a characteristic of these inorganic ferments also.[2] The failure of the older observers to discover that the reactions catalyzed by the metallic colloids are incomplete is to be explained by the character of the reactions which they studied. In the catalysis of hydrogen peroxide, for example, by platinumsol, one of the products of the decomposition—the oxygen—is allowed to escape as soon as liberated, so that even if water could be oxidized to hydrogen peroxide it would not occur under the conditions of the experiment. For, in order that reversion may occur, all the products of the decomposition must be allowed to accumulate. If the products of a chemical reaction catalyzed by any ferment are removed as fast as formed, that reaction is complete and shows no "limit."

A great analogy exists also between the sensitiveness of the true ferments to certain poisons and the sensitiveness of the colloidal solutions of the noble metals to these same poisons. Acids, bases, and salts affect different organic ferments in different ways. The effect of these same substances upon the inorganic ferments is equally marked. The addition of a mere trace of disodium phosphate to a liter of colloidal platinum solution caused an immediate fall in the "velocity constant" of the decomposition of hydrogen peroxide from 0.023 to 0.015. After waiting several

---

[1] Neilson: Science, 1902, XV, p. 715.

[2] The spongy platinum used by Neilson is not a colloid, but the nature of its action is no doubt the same as that of platinumsol.

days the velocity constant had fallen lower still—to 0.011. We are familiar with this same inactivation in the case of the true ferments, and it is not impossible that that which causes the platinumsol to become inactive, namely, a precipitation of the colloidal par icles, lies at the basis of the inactivation of the true ferments also.

Poisons, the action of which upon organic ferments is more or less characteristic and striking, show this same action when brought in contact with inorganic ferments. The physiological action of hydrocyanic acid finds its explanation in its power to interfere with the oxidizing ferments of the cell. When added in even exceedingly minute traces to oxidizing (and other) ferments it reduces their action to a point where it can scarcely be recognized. The addition of hydrocyanic acid to platinumsol reduces its action upon hydrogen peroxide in the same striking way, for the presence of 0.0014 milligram hydrocyanic acid in a liter of the colloidal platinum solution reduces its action one-half. Hydrogen sulphide, which also has a powerful action in inhibiting the activity of organic ferments, shows the same behavior when added to the colloidal solutions of the noble metals. The order in which the poisons, ferments, and hydrogen peroxide are put together is not without influence upon the extent of the inhibition produced. The fact is therefore of interest that the order which is most effective in reducing the action of an organic ferment is also the most effective when an inorganic ferment is dealt with.

CHAPTER V.

THE ACTION OF THE ENZYMES FOUND IN THE HUMAN ALIMENTARY TRACT.

1. Amylase (ptyalin, amylopsin, diastase) is the term applied to the enzyme found in the salivary and pancreatic secretions, which has the power of acting upon starch and producing from it maltose and dextrin. The ferment is found in other tissues of the body also, as well as in germinating cereals of all kinds (rice, corn, oats, etc.). The alimentary secretions of all animals do not contain amylase. The saliva of the dog, for instance, contains no starch-splitting ferment, and the same seems to be true of all carnivora. The amylase of the parotid seems to be present in the human being immediately after birth, but the starch-splitting ferment of the pancreas does not appear until a month or more later.

We recognize the presence of amylase by its power of acting upon starch and its ready destruction by heat rather than by any analytical means at our disposal. As with practically all enzymes, amylase has not yet been obtained in a pure state. COHNHEIM[1] has succeeded in obtaining the amylase of the saliva as a gray powder of fairly constant composition by acidifying the collected saliva with phosphoric acid and neutralizing subsequently with dilute calcium hydroxide. The amylase is carried down mechanically with the precipitate of calcium phosphate, and after filtering is redissolved by the addition of water. From this aqueous solution the amylase can be reprecipitated by alcohol in order to further purify it.

---

[1] COHNHEIM: Virchow's Archiv, 1863, XXVIII, p. 241.

The amylase of the pancreas is obtained in a very active state by ROBERT's method of precipitation with alcohol and re-solution in water. The method of COHNHEIM for obtaining the starch-splitting ferment of the saliva has also been applied to the pancreas, though ROBERT's method seems to give a more active preparation.

The most frequently utilized source of amylase to-day is probably germinating malt, which contains this enzyme in enormous quantities. EFFRONT's [1] method of obtaining the ferment from this source consists in the extraction of finely ground malt with water for some time, filtration, and alcoholic fermentation of the filtrate through the addition of yeast. After a second filtration the amylase is precipitated from the filtrate through the addition of alcohol. EFFRONT obtained in this way from every 100 grams of malt 3 to 3½ grams of a white substance which, when redissolved in water, was as active as 80 grams of the original malt.

Much discussion has arisen as to the identity or non-identity of the amylase obtained from various sources. So far as the *qualitative* character of the action of the amylase on starch is concerned, there seems to be no difference, whatever be the origin of the ferment. *Quantitatively* the pancreatic amylase acts more powerfully than the salivary, but this is explicable on the basis of mere differences in the concentration of the enzyme in the two secretions.

Other characteristics, such as differences in sensitiveness to heat, alkalies, acids, salts, etc., have also been brought forward in support of the independent nature of the starch-splitting ferments from different sources, but great care must be exercised in accepting these arguments. As already stated, amylase has not yet been obtained in a pure state. The impurities which accompany any preparation of amylase are different, depending upon the source of the ferment. Even

---

[1] EFFRONT: Die Diastasen. Translated into German by BÜCHELER, Leipzig u. Wien. 1900, I, p. 113.

the same source does not always yield the same impurities. It is therefore entirely probable that the differences which have been assumed to exist in the amylase obtained from different sources are only apparent, and depend upon the fact that the impurities accompanying the ferment are different. In this way the same ferment is simply compelled to work under different external conditions, and, as will be shown immediately, the medium in which amylase acts has an important influence upon its action. What has been said here of amylase holds also for the other ferments.

Amylase brings about two changes when allowed to act upon starch—it liquefies the starch and converts it into sugar. The sugar which is formed by amylase when unmixed with other enzymes is the disaccharide, maltose. When amylase is allowed to act upon raw starch granules, the latter are seen to be gradually eroded, and sugar is formed. Upon boiled starch amylase acts far more energetically, the conversion into sugar being accomplished in much shorter time. According to KÜHNE, the amylase of the saliva does not act upon unboiled starches at all, while that of the pancreas does. It is entirely probable, however, that this difference is only apparent and dependent upon the fact that the concentration of the amylase in the saliva is less than that in the pancreatic juice. The chemical change which starch suffers under the influence of amylase is indicated in the following formula:

$$C_{12}H_{20}O_{10} + H_2O = C_{12}H_{22}O_{11}.$$
$$\text{Starch} \qquad \text{Water} \qquad \text{Maltose}$$

This formula shows that starch undergoes a hydration when acted upon by amylase, but it tells us nothing of the mechanism of this change. That maltose is the ultimate product of the catalytic change is almost universally accepted, but opinions differ widely as to the intermediate changes which occur in the starch before the final stage of maltose is reached. All observers are agreed, however, that dextrin is formed in the process of saccharification. Whether the dex-

trin is formed from the starch and is subsequently changed
into sugar, or whether dextrin and sugar are formed simultaneously, is not yet settled.   If the former is the correct
view, it would be expressed by the formula

$$C_{12}H_{20}O_{10} = C_{12}H_{20}O_{10}. \qquad (1)$$
$$\text{Starch} \qquad\qquad \text{Dextrin}$$

$$C_{12}H_{20}O_{10} + H_2O = C_{12}H_{22}O_{11}. \qquad (2)$$
$$\text{Dextrin} \qquad \text{Water} \qquad \text{Maltose}$$

If the second theory, that of MUSCULUS, is the correct one,
the following equation would hold:

$$2\,C_{12}H_{20}O_{10} + H_2O = C_{12}H_{22}O_{11} + C_{12}H_{20}O_{10}.$$
$$\text{Starch} \qquad \text{Water} \qquad \text{Maltose} \qquad \text{Dextrin}$$

The majority of authors also believe that the dextrin formed
is not all of the same kind, that, in other words, *several* dextrins are formed, which have been distinguished by different
names.   The basis for this belief lies in the fact that the
dextrins obtained at different stages in the saccharification
of starch are not converted into maltose with the same rapidity
when acted upon by amylase under normal circumstances
or in the presence of acids, salts, various organic substances,
etc.   It seems probable, however, that these differences rest
more upon physical than upon chemical characteristics of
the dextrin.   Dextrin belongs to the group of substances
known as colloids,[1] and the properties which have been
considered characteristic of different dextrins might readily
characterize one and the same chemical substance in different physical states.   A dextrin solution in which the particles
are finely divided would, for example, be more rapidly converted into sugar than one in which the suspended particles
are coarser.

Amylase may be used as a type to illustrate some of the
characteristics of a ferment.   Amylase accelerates a chemical

---

[1] See Chapter XIV, Part 2.

reaction which occurs also in its absence, only then exceed. ingly slowly. Starch paste by itself kept at a suitable temperature is very slowly converted into maltose. Amylase is therefore a positive catalyzer. The influence of temperature upon the activity of this ferment has been worked out by KJELDAHL. At 0° C. amylase acts exceedingly slowly, but its activity increases rapidly with an increase in temperature until 35° C. is reached. From 35° to a little above 60° C. a further but less marked increase is observed. Beyond this point the activity of the ferment falls off rapidly until at 70° C. it acts no more energetically than at about 15°. A curve similar to those shown on p. 89 would therefore represent graphically the influence of temperature upon amylase.[1]

The reaction which amylase catalyzes is incomplete when allowed to take place in glass, the saccharification coming to a standstill when less than half of the total starch present has been converted into maltose. The exact point varies with external conditions of temperature, concentration, etc. We shall see later that this fact indicates that the action of amylase is reversible. Evidence of the truth of this statement has recently been brought by CREMER,[2] who has shown that a ferment (perhaps identical with amylase) which has the power of splitting glycogen (an isomer of starch, and often called animal starch) into glucose is also able to synthesize glycogen from glucose.

If sufficient time is given, the amount of change brought about in a starch paste is the same, no matter what the concentration of the amylase.

The activity of amylase is markedly influenced by the presence of different chemical substances in the reaction mixture. Amylase seems to act best in a neutral medium.

---

[1] See EFFRONT: Die Diastasen. Translated into German by BÜCHELER, Leipzig, 1900, I, p. 119.

[2] CREMER: Ergebnisse d. Physiol., 1902, I, 1te Abth., p. 803.

Both acids and alkalies soon bring its activity to a stand-still, even in very moderate concentrations. The influence of acids upon amylolytic activity is of physiological importance, because the chewed food mixed with saliva passes into the stomach. The concentration of the acid in the stomach is sufficient to stop the activity of the amylase; but, as will be remembered from what has been said before, the amylase continues its action for some time, especially in the cardiac end of the stomach, as the food which passes into this viscus is not at once rendered acid by its secretions.

It has been shown by the work of CANNON that the saliva does not act best in the concentration in which it is poured out upon the food, but when diluted with about three times its bulk of water. The explanation of this fact lies no doubt in the dilution of the products of amylolytic activity, for in a concentrated solution the point at which the reaction comes to a standstill is reached sooner than in a more dilute one. In fact, most fermentation mixtures which have come to a standstill will go on further if only water is added. The use of water with meals, after thorough mastication of the food, is therefore not only not harmful, but useful. Dilution of the stomach contents hastens the activities of the enzymes here also. The presence of water in the stomach, moreover, increases the secretion of gastric juice and hastens absorption. The evil consequences of the consumption of water with meals lie in its use before mastication of the food, thus preventing thorough insalivation and the good effects of this process upon the food; and in the use of too *cold* water, whereby the reaction velocity of all fermentative changes is markedly decreased.

2. **Maltase** (glucase) is an enzyme the characteristic activity of which lies in its power of splitting maltose into two molecules of dextrose. It is able to act also upon starches and dextrins, but only exceedingly weakly when compared with amylase, the ultimate product from these substances

also being dextrose. Maltose is formed as an intermediate product, so that what was defined as the characteristic activity of maltase appears here also. It is evident from this that care must be exercised in deciding whether a ferment present in animal or vegetable tissues and capable of producing a sugar from starch is amylase or maltase, or whether both are present. The recognition of the first is dependent upon the identification of maltose as the end-product of the reaction, that of the latter upon the recognition of dextrose. The existence of maltase in the presence of amylase can be proved by isolating a ferment which splits pure maltose into dextrose. The recognition of amylase in the presence of maltase is more difficult, and differences in resistance to heat, chemical reagents, etc., must be utilized to bring about their separation.

Maltase seems to have been first discovered by BECHAMP in the urine, and somewhat later by BROWN and HERON in the pancreatic juice and in the small intestine. It is found also in the saliva. The commonest source of maltase to-day for the study of its activities is corn, from which it can be obtained in a very active state, by precipitation of a dilute acid-alcohol extract of this cereal with strong alcohol.[1] The ferment when obtained in this way is not pure.

Qualitatively the maltase obtained from different sources shows the same action upon maltose. Considerable differences exist, however, in the resistance of this enzyme to external conditions when obtained from different animal tissues or cereals. This has been taken to argue in favor of the existence of different maltases. Since maltase has not yet been obtained in a pure state, however, we cannot be sure whether the differences found in the various maltase preparations are due to specific differences in the ferments themselves or to differences in the impurities with which the preparations are contaminated.

---

[1] BEIJERINCK: Centralbl. f. Bakteriologie, 1898, xxiii, 2te Abth.

LINTNER and KROEBER [1] give as the optimal temperature for the action of yeast maltase upon maltose 40° C. Below and above this point the formation of dextrose proceeds much more slowly. At 50° C. this maltase is almost entirely destroyed. CUSENIER gives 56° to 60° C. as the optimal temperature for the maltase obtained from corn.[2]

Maltase is of special biological interest, as it was in a study of this ferment that the fundamental fact of the reversible action of enzymes was first experimentally proved by the interesting work of A. CROFT HILL.[3] For a better understanding of this property of enzymes a few introductory words are necessary to illustrate the general conceptions of *chemical equilibrium* and *reversion*.

If a number of substances capable of reacting chemically with each other are brought together, a reaction ensues, which after a time comes to a standstill (practically speaking). We say then that the system is in chemical equilibrium. If, for example, at a definite temperature chemically equivalent amounts of acetic acid and ethyl alcohol are mixed together, a reaction ensues according to the following equation:

$$CH_3COOH + C_2H_5OH = CH_3COOC_2H_5 + H_2O.$$

Acetic acid   Ethyl alcohol   Ethyl acetate   Water

The reaction takes place in the direction from left to right. If chemically equivalent amounts of ethyl acetate and water are mixed together, ethyl alcohol and acetic acid are formed. In other words, the above reaction takes place from right to left. Neither in the first nor second instance does the reaction become complete. Before the given amounts of

[1] LINTNER and KROEBER: Berichte d. deutschen chem. Gesellsch., 1895, p. 1050.

[2] Quoted from EFFRONT: Die Diastasen. Translated into German by BÜCHELER, Leipzig u. Wien, 1900, I, p. 225.

[3] HILL: Jour. of the Chem. Society, 1898, LXXIII, p. 634; Berichte d. deutschen chem. Gesellsch., 1901. XXIV, p. 1380.

acetic acid and ethyl alcohol, or ethyl acetate and water, have undergone complete decomposition the reaction comes to a standstill.

A reaction such as we have just spoken of, which can take place from right to left as well as from left to right, is called a *reversible reaction.* We indicate such a reversible reaction as follows:

$$CH_3COOH + C_2H_5OH \rightleftarrows CH_3COOC_2H_5 + H_2O.$$

Acetic acid   Ethyl   Ethyl acetate   Water
            alcohol

It can readily be seen that when equilibrium is established in a reversible reaction the four substances reacting with each other are present in the reaction mixture. The characteristic feature of such a condition of equilibrium is found in the fact that under the same external conditions it is always the same no matter from which side it is reached. In other words, it is immaterial whether chemically equivalent amounts of acetic acid and ethyl alcohol or chemically equivalent amounts of ethyl acetate and water are mixed together. The condition of equilibrium reached in either case is the same.[1]

Although we say ordinarily that when chemical equilibrium has been established the reaction has come to a standstill, this is in reality incorrect. When chemical equilibrium is established between the two members of a chemical equation it really means that the chemical changes are still going on, only the amount of change in the one direction is exactly counterbalanced by the reverse change in the opposite direction. The reaction is therefore *stationary.*

As has long been known, maltase acting upon maltose is unable to bring about its complete analysis into dextrose. The reaction comes to a standstill when some 85 percent of the original maltose is split. HILL now tried the interesting

---

[1] See COHEN Physical Chemistry for Physicians. Translated by MARTIN H. FISCHER, New York, 1903, p. 68.

experiment of allowing yeast maltase to act at 30° C. upon a 40-percent solution of dextrose for a number of months, and found that this same ferment, which is able to bring about the analysis of the maltose, is also able to bring about its synthesis from the products of the analysis. HILL determined the appearance of maltose and its gradual increase in the dextrose solution by means of the polariscope and copper reduction tests. Since maltose is able to rotate the plane of polarized light more to the right than dextrose, and since dextrose has a greater reducing power than maltose, it is possible through analysis to determine whether in a mixture of the two sugars one is increasing at the expense of the other. Column A in the following table shows how, under the influence of yeast maltase, the rotating power of a pure dextrose solution gradually increases, while column B shows how its reducing power gradually decreases. Columns C and D indicate the percent of maltose contained in the originally pure dextrose solution, as calculated from the figures obtained in the first two columns. The values obtained by the two methods of analysis agree very well.

Duration of experiment	A. Rotation.	B. Reduction.	C. Maltose in %.	D. Maltose in %.
0 days	52.5°	100.5	0.0	0.0
5 "	55.0°	97.7	3.0	3.5
14 "	58.3°	95.8	7.4	6.8
28 "	60.3°	94.0	10.0	10.0
42 "	62.7°	92.5	13.0	12.0
70 "	63.6°	90.6	14.0	15.0

Maltase seems, therefore, to have a reversible action and is able not only to split maltose into dextrose, but also to synthesize the disaccharide from the monosaccharide. The chemical reaction expressing this may be written as follows:

$$C_{12}H_{22}O_{11} + H_2O \rightleftarrows C_6H_{12}O_6 + C_6H_{12}O_6.$$

Maltose   Water   Dextrose   Dextrose

This equation is entirely analogous to the one given on page 107. As in the case of ethyl acetate and water, we have here also a reaction which can take place in either direction. No matter from which side we begin, the final result is always the same—the presence in the reaction mixture of maltose and water side by side with two molecules of dextrose. The only difference between the reaction discussed on page 107 and this one is that the analysis or synthesis of maltose occurs under the influence of a catalyzer. But, as has been pointed out, a catalyzer only alters the rate of a chemical reaction and does not enter into the reaction itself (unless it does so temporarily). The analysis or synthesis of maltose would occur in the same way, and the state of equilibrium finally attained be the same, even if no maltase were present, only under such circumstances the velocity of the reaction would be very low.

The system

$$\text{Maltose} + \text{Water} \rightleftarrows \text{Dextrose} + \text{Dextrose}$$

is in equilibrium under the conditions existing in HILL's experiments when some 85 percent of dextrose exists beside about 15 percent of maltose. It is clear that such a mixture of the two sugars in water will undergo no apparent change whether maltase be present or not. As shown above, however, the reaction has not ceased, only the chemical change in the one direction just counterbalances that in the opposite, so that outwardly everything remains the same. But what must occur if into the above reaction mixture some maltose or some dextrose is introduced, or either of these substances is removed? It is clear that this must disturb the chemical equilibrium existing between the maltose on the one hand and the dextrose on the other. Whenever this occurs the activity of maltase as an agent hastening the establishment of a state of equilibrium between these two substances must come into play, and must either analyze or synthesize maltose until such a state of equilibrium is again established. The introduction of dextrose or the removal of maltose from a reaction mixture containing the two sugars in chemical equilibrium would therefore be followed by a synthesis of maltose, while the introduction of maltose or the removal of dextrose would be followed by an analysis of the disaccharide if other external conditions remained the same.

3. **Caseinase** (rennin, rennet) is the name given to a ferment found in many animal and vegetable cells and their secretions, which has the power of changing caseinogen into casein—that is, the power of curdling milk.   In the human being the ferment is found in the secretions of the stomach from birth. Milk-curdling ferment is found also in the pancreatic juice, and though it differs somewhat from the caseinase found in the stomach in its resistance to heat and chemical agents, it is probably the same substance and may well be called by the same name.

A number of authors have pointed out the fact that caseinase always coexists with a proteolytic ferment.   In the case of the human being (and certain other animals) caseinase accompanies the acid-proteinase (pepsin) of the stomach and the alkali-proteinase (trypsin) of the pancreas.   Caseinase, moreover, is found in largest amount in those portions of the stomach where the most acid-proteinase is found, namely, the fundus.   The constant association of the two enzymes has given rise to the idea that they probably represent a giant molecule, the different portions of which have different fermentative functions and are differently affected by external conditions.   It has been shown by HERZOG, for example, that antiproteinase (antipepsin) added to the combined ferments inhibits only the proteolytic action, while the milk-curdling action goes on undisturbedly.   Digestion with weak acids will at proper temperature quickly destroy the caseinase constituent of the entire molecule, and by mechanical precipitation HAMMARSTEN has succeeded in obtaining a caseinase solution free from any proteolytic activity.

Preparations of caseinase are obtained by extracting, with

solvents of various kinds, the stomach of the calf, which contains the enzyme in enormous quantities. Among these sodium chloride and hydrochloric acid solutions may be mentioned. Glycerine is also used extensively. When purer preparations are required, precipitation with alcohol, filtration, and re-solution in water are necessary.

Caseinase behaves like other ferments in that it is readily destroyed at a rather low temperature. A few minutes' exposure in a neutral solution at 70° C., or in an acid solution at 63° C., suffices to do away with all milk-curdling properties. In an alkaline medium the ferment is affected rapidly even at a very low (room) temperature. Though active in a faintly acid or alkaline medium, caseinase acts best in a neutral medium and at a temperature of about 40°. Below and above this point the rapidity with which milk is made to coagulate falls off rapidly.

The mechanism of the coagulation of milk has been studied by a number of authors. Briefly summarized it may be stated as follows: Milk contains a protein substance to which the name *caseinogen* has been given (casein of HAMMARSTEN), the change of which into *casein* (paracasein of HAMMARSTEN) is the essential change in the curdling of milk. Casein is formed from caseinogen under a number of circumstances. This occurs when acid is added to the milk, or when any acid-producing change occurs in the milk such as "souring" under the influence of bacteria, or when an electric current is passed through the milk. But milk is caused to coagulate most rapidly in ordinary alimentation through the addition of caseinase. In the gastro-intestinal tract two agencies are active in bringing about the coagulation of ingested milk: first, the hydrochloric acid of the gastric juice, and secondly, the presence of caseinase in the gastric and pancreatic secretions. Another source of caseinase in the alimentary tract is found at times in certain bacteria which may be present.

When caseinase is added to milk the latter soon sets into

a solid mass which contracts after a time and squeezes out a yellowish fluid—the *whey*. The clot consists of casein in the meshes of which there is found the fat of the milk. The whey contains lactalbumin and lactglobulin, two protein bodies which have nothing to do with the curdling of the milk, together with the sugar of the milk and most of its salts.

In the coagulation of milk during ordinary digestion three substances are involved—caseinogen, caseinase, and certain salts. It was first shown by HAMMARSTEN that calcium salts play an important rôle in the curdling of milk. As will be shown immediately, however, calcium represents only one of a large number of salts which can influence the coagulation of milk. Various theories have been proposed from time to time to explain the interaction of the three substances. These have in the main been of a chemical nature, and have assumed the production of true chemical combinations between caseinogen and calcium, etc. The recent work of CONRADI and of LOEVENHART speak against this idea and indicate that *in the curdling of milk we are dealing with a physical process in which the colloid caseinogen which exists in ordinary milk in a liquid or sol state is converted into the solid or gel state which we term casein.* Caseinogen and casein (casein and paracasein of HAMMARSTEN) are therefore the same chemically, but different physically in that the former consists of small particles suspended in the liquid portion of the milk, while the latter represents these same particles clumped. The passage from the sol to the gel state is influenced by the ferment caseinase and certain salts.

Casein can be precipitated by all methods used for this purpose much more easily than can caseinogen. But if the proper concentration is employed the latter is brought down also. It is concluded from this that casein exists in a coarser state of aggregation than caseinogen. Nor can the differences which have been said to exist between caseinogen and casein be taken to indicate that they are not the same chemically, for all the stated differences can easily be explained on the assumption that the two represent physical modifications of one and the same substance.

As stated above, calcium salts are not .the only ones to bring about a precipitation of casein. At the proper concentration a large number of salts cause the precipitation not only of casein but of caseinogen as well. Arranged in the order of their effectiveness the salts of the different metals read as follows:

(1) Metals which precipitate neither casein nor caseinogen: Sodium, potassium, ammonium, rubidium (?), cæsium (?).

(2) Metals which at room temperature quickly precipitate casein, but caseinogen only after remaining at 40° C. for some time or on heating to a still higher temperature: Lithium, beryllium, magnesium, calcium, strontium, barium, manganese (ous), iron (ous), cobalt (ous), nickel (ous).

(3) Metals which precipitate both casein and caseinogen at room temperature: The remaining heavy metals, including iron (ic).

As we pass from the first group through the second to the third, the precipitating power of the metals for both casein and caseinogen increases progressively.[1]

---

[1] See OSBORNE: Journal of Physiology, 1901, XXVII, p. 398; CONRADI: Münchener medizinalische Wochenschr. XLVIII, 1901, (1), p. 175; LOEVENHART: Zeitschrift für physiologische Chemie, 1904, XLI, p. 177.

CHAPTER VI.

THE ACTION OF THE ENZYMES FOUND IN THE HUMAN
ALIMENTARY TRACT (*Continued*).

4. Acid-proteinase (pepsin).—By this term is understood
a proteolytic ferment which acts only in the presence of
an acid. Acid-proteinase is found widely distributed in
nature in both animal and vegetable cells. It is found in
many regions in the human body, more especially in the
muscles, and, what interests us most in this connection, in
the mucous lining and secretion of the stomach. Acid-
proteinase has the power of acting upon proteins in the
presence of certain acids and splitting these proteins into
a series of simpler substances. What these simpler substances are will be discussed further on.

Acid-proteinase is present in the stomach of the human
being from the time of birth. Neither at this time nor later
in life, however, do all parts of the stomach contain it in
the same amount. The cardiac end of the human stomach
contains much more than the pyloric end.

Pure preparations of acid-proteinase are exceedingly difficult to obtain. Extraction of the mucosa of the pig's stomach
with a 0.2 percent hydrochloric acid solution or with glycerine
yields very active preparations. These are, however, very
impure. BRÜCKE's method of extraction of the mucous
membrane with dilute phosphoric acid and subsequent
neutralization with calcium hydroxide seems to yield very
pure acid-proteinase. The enzyme is carried down mechanically with the precipitate of calcium phosphate, and can be

114

subsequently redissolved in water. PEKELHARING[1] claims to have prepared the purest acid-proteinase thus far obtained and classes the enzyme among the proteins. His preparation gives the well-known reactions for proteins, and on quantitative analysis shows the presence of carbon, nitrogen, hydrogen, and sulphur in the proportions in which they exist in these bodies. As the preparation contains no phosphorus it is not considered a nucleo-proteid. In harmony with the findings of NENCKI and SIEBER, PEKELHARING looks upon chlorine as a constant constituent of acid-proteinase. Should it be shown ultimately that PEKELHARING's enzyme is not a pure substance it will still have to be looked upon as the most active preparation of this ferment obtained thus far. 0.001 milligram in 6 c.c. of a 0.2 percent hydrochloric acid solution dissolved a flake of fibrin in a few hours.

The acid-proteinase of PEKELHARING is able not only to act on proteins but also curdles milk. This means that acid-proteinase and caseinase are probably parts of the same molecule. As stated above, gastric juice contains a fat-splitting ferment—lipase—but the pepsin preparation of PEKELHARING shows no fat-splitting activity. This author's work, therefore, supports a view which has been suggested by NENCKI and SIEBER before him, that acid-proteinase and caseinase represent different parts of a giant molecule.

BRUNTON,[2] and FRIEDENTHAL and MIYAMOTA[3] have expressed themselves as opposed to the conception of the protein character of pure acid-proteinase and bring forward as proof that they have succeeded in obtaining active acid-proteinase preparations which do not give any of the reactions for proteins. It is also pointed out by these authors that acid-proteinase is not digested by alkali-proteinase (trypsin), which as

---

[1] PEKELHARING: Zeitschrift für physiologische Chemie, 1902, XXXV, p. 8.

[2] BRUNTON. Centralblatt für Physiologie, 1902, XVI, p. 201.

[3] FRIEDENTHAL and MIYAMOTA: Centralblatt für Physiologie 1901, XV, p. 786; ibid., 1902, XVI, p. 1.

is well known acts upon all proteins thus far studied. Against FRIEDENTHAL and MIYAMOTA can be brought the argument that their preparations were not very active and so may have contained too little of the pure enzyme to give the protein reactions. Further study of the subject is, however, necessary before the question can be looked upon as settled.

Acid-proteinase acts only in an acid medium, but the kind of acid is not immaterial. Hydrochloric acid furnishes by far the most favorable medium, but nitric, sulphuric, phosphoric, lactic, acetic, tartaric, etc., are also active. PFLEIDERER,[1] who has investigated the question physico-chemically, comes to the conclusion that, broadly speaking, different acids favor the action of acid-proteinase on fibrin, according to their "avidity," those with the greater "avidity" acting more powerfully than those with a lesser one. Since we now know that the "avidity" or "strength" of an acid is an expression of the number of free hydrogen ions it yields upon solution, the above statement may be said to mean that those acids which yield the largest number of hydrogen ions when dissolved in water are most powerful in furthering the activity of acid-proteinase. The following table may serve to illustrate what has been said as well as serve as a text for that which is to follow. The degree of fibrin digesion as measured by GRÜTZNER's method[2] is expressed in

Name of acid, all ⅒ normal.	Degree of color after									
	5 min.	10 min.	20 min.	30 min.	1 hr.	1½ hrs.	2 hrs.	5 hrs.	10 hrs.	24 hrs.
Hydrochloric	1	5	5–6	6	6	6	6	6	6	6
Nitric	0	0–1	1	2	3–4	5	5	5–6	6	6
Phosphoric	0	0	0–1	1	3–4	5–6	6	6	6	6
Lactic	0	0	0	0	0–1	1	2–3	4	5–6	6
Acetic	0	0	0	0	0	0	0	0	0	0
Sulphuric	0	0	0	0	0	0	0	0	0	0

[1] PFLEIDERER: PFLÜGER'S Arch., 1897, LXVI, p. 605.
[2] See p. 130.

figures from 1 to 6, according to the amount of color produced. In each digestion-tube were present the same amounts of fibrin and pepsin and chemically equivalent amounts of the various acids.

At a somewhat higher concentration (for example, 1/20 normal) both the acetic and sulphuric acid tubes would have shown some degree of fibrin digestion, but the order of the table would not have been much different. Without further comment, therefore, it is clear that while the effect of any acid upon the proteolytic activity of acid-proteinase is chiefly a function of the number of hydrogen ions which the acid yields on solution in water, the individual acids vary enough from each other even when the number of hydrogen ions in the unit volume of the digestion mixture is the same to make us inquire after the cause of this difference. Sulphuric acid, for example, which in dilute solution is dissociated to about the same degree as hydrochloric, stands far below this in its power of bringing about a demolition of the protein molecule. Now since sulphates in general (sodium sulphate, magnesium sulphate, etc.) markedly retard the activity of acid-proteinase even when this is acting in the presence of hydrochloric acid, it lies near at hand to consider the SO_4 constituent of the sulphuric acid and of these salts as chiefly responsible. Other salts, such as the chlorides and iodides of sodium, potassium, and ammonium, also inhibit the digestion of a protein under the influence of acid proteinase and hydrochloric acid, but not so markedly as the sulphates.

The reason for the individual differences between the acids and the effect of various salts upon the velocity of digestion is not yet entirely clear. It is evident that the nature of their action might be various. In looking over PFLEIDERER's tables, however, one is struck with the parallelism which exists everywhere between the effect of the different acids and the various salts upon the rate of digestion and the degree of swelling which these substances bring about in fibrin alone. As is well known, fibrin swells in different acids, but not to the same degree in each. If now the acids are arranged according to the degree in which they make fibrin take up water (swell), this order is the same as that given above for the effect of these same acids on digestion under the influence of acid-proteinase. Sulphuric acid, for example, which stands very low in the list of acids as favoring acid-proteinase digestion, occupies a similar position when the acids are arranged in the order in which they cause fibrin to swell. What has been said of the acids holds also for the salts. The more a salt inhibits the absorption of water by fibrin (for example, a sulphate) the more does this salt retard the digestion of a protein.

The effect of various external conditions upon the proteolytic activity of acid-proteinase has usually been attributed to the effect of these conditions upon the acid-proteinase itself. Unquestionably external conditions can most markedly influence the state of the ferment itself—it is no doubt a colloid and influenced, as are all colloids, by external conditions—but in the experiments which have been cited, the chief effect of the external conditions seems to have been on the protein undergoing digestion. BRÜCKE many years ago pointed out that the more fibrin has swelled the more rapidly it is digested. The different acids and salts influence this swelling in different degrees, in consequence of which the protein is attacked by the acid-proteinase with greater or less ease. It seems to me that this difference between the *physical* change which a protein suffers when it swells during the process of peptic digestion, and the *chemical* change which probably constitutes the real activity of the proteolytic ferment when the simpler digestion-products are formed, has never been sufficiently well drawn. The two are different, and differently influenced by external conditions.

The optimum concentration of hydrochloric acid for the activity of acid-proteinase varies with the protein which is being acted upon. HAMMARSTEN gives 0.08 to 0.1 percent for fibrin; 0.1 percent for myosin, casein, and vegetable proteins; 0.25 percent for coagulated white of egg.

Alkalies and alkaline salts when present in a digestion mixture are uniformly injurious in their action. If proteins are present at the same time with the acid-proteinase, alkalies act less destructively, no doubt because the alkalies combine with the protein. This is indicated by the fact that the same concentration of alkali acts the less deleteriously upon the acid-proteinase the greater the amount of protein present.[1] Of other substances which, when present, influence the activity of acid-proteinase and which are of medical interest, the following may be mentioned: Alcohol and tannic acid interfere with the action of the ferment, as do most alkaloids and carbohydrates. Bile also belongs in this group, for one part of this substance is able to do away entirely with the proteolytic activity of five hundred parts of gastric juice. Caffein and theobromin further the action of the enzyme.[2]

Within certain limits the rapidity with which acid-proteinase splits a protein increases with an increase in the quantity of the ferment present in the reaction mixture. But in no case is the digestion of the protein complete unless the products of the reaction are removed. An accumulation of the products of digestion retards markedly the further analysis of the protein. It will be shown further on that this is probably dependent upon the fact that the action of acid-proteinase is reversible, and that the ferment synthesizes from the products of the digestion the protein itself,

[1] See LANGLEY. Journal of Physiology, 1882, III, pp. 253 and 283; LANGLEY and EDKINS: ibid., 1886, VII, p. 371.

[2] WROBLEWSKI: Zeitschrift für physiologische Chemie, 1895, XXI, p. 1; also BUCHNER. Berichte der deutschen chem. Gesellschaft, 1897, XXXIII. p. 1110; LABORDE: Comptes rendus Soc. biol., 1899, LI, p. 821; NIRENSTEIN and SCHIFF. Archiv für Verdauungskrankheiten, 1902, VIII, p. 559. BRUNO. PAWLOW's Work of the Digestive Glands. Translated by THOMPSON London, 1902, p. 158.

in a way analogous to that described for the action of mal
tase on maltose.

Acid-proteinase is, like other ferments, very sensitive t
temperature. At 0° C. it is scarcely active. From this poin
to 35° C. the velocity of its action increases progressively
attaining an optimum between 35° and 50° C. Beyond thi
point its activity diminishes.[1] In a neutral solution the fer
ment is in a few minutes destroyed at 55° C. The presenc
of hydrochloric acid protects the ferment against destructio1
by heat, so that when present in the concentration in whicl
the acid is found in the stomach, the ferment is destroyed onl}
slowly until 65° C. is reached. Peptones and certain salts als
have a protecting influence.[2] As with other ferments, th
concentration of the acid-proteinase itself in a pure solutio
influences the temperature at which it is most rapidl
destroyed.

When natural or artificial gastric juice (a solution of acid
proteinase in dilute hydrochloric acid) is allowed to act upon
a protein, such as fibrin, the following changes are observed.
The fibrin begins to swell and its superficial portions become
translucent, until, under favorable conditions of temperature,
the insoluble protein is rendered soluble. If given time
enough, the acid alone will bring about a solution of the fibrin;
but if acid-proteinase is present this change occurs much more
rapidly. If the enzyme is employed in a neutral solution the
protein is not dissolved. Acid and ferment together are
therefore necessary to bring about the rapid change from the
protein to the soluble product. The soluble product consists,
as will be seen presently, of a mixture of a number of chemical
substances, which are distinguished from the original protein
upon which the acid-proteinase acted by their ready solubility
in water and their ready diffusibility through animal and
vegetable membranes. These properties, it will be noticed,

---

[1] v. WITTICH: PFLÜGER's Arch., 1869, II, p. 193; ibid., 1870, III,
p. 339.

[2] BIERNACKI. Zeitschr. für Biol., 1892, XXVIII, p. 453.

stand in marked contrast to those of·the original protein substance—for instance, fibrin—which is insoluble and readily held back by a filter or animal membrane. It is these properties of the products of gastric digestion which give them the power of being easily taken up by the mucous membrane of the alimentary tract.

Our conceptions of *the nature of the products formed when proteins are digested in the presence of acid-proteinase* have, within the last few years, been markedly altered, more especially through the investigations of ZUNZ,[1] PFAUNDLER,[2] LAWROW,[3] MALFATTI,[4] SALASKIN,[5] LANGSTEIN,[6] EMIL FISCHER [7] and ABDERHALDEN.[7] The work of all these authors indicates that acid-proteinase working in the presence of an acid causes a cleavage of the protein molecule into substances of a much simpler chemical composition than we formerly supposed. Where we once believed that proteoses and "peptones," in the sense in which KÜHNE used this term, constituted the final products of gastric digestion, we now know that a large number of substances, which were formerly looked upon as produced only in pancreatic digestion, are also formed. In experiments in which acid and acid-proteinase have been allowed to act long enough, there have been found, besides the more complex proteoses and "peptones," the following simple compounds: leucin, tyrosin, alanin, phenylalanin, amino-

---

[1] ZUNZ: Zeitschrift für physiologische Chemie, 1899, XXVIII, p. 132.

[2] PFAUNDLER. Zeitschrift für physiologische Chemie, 1900, XXX, p. 90.

[3] LAWROW: Zeitschrift für physiologische Chemie, 1899, XXVI, p. 513; ibid., 1901, XXXIII, p. 312.

[4] MALFATTI: Zeitschrift für physiologische Chemie, 1900, XXXI, p. 43.

[5] SALASKIN: Zeitschrift für physiologische Chemie, 1901, XXXII, p. 592.

[6] LANGSTEIN: HOFMEISTER's Beiträge zur chemischen Physiologie, 1901, I, p. 507.

[7] EMIL FISCHER and ABDERHALDEN: Zeitschrift für physiologische Chemie, 1903, XXXIX, p. 81.

valerianic acid, aspartic. acid, glutamic acid, and lysin. It will be seen later that these compounds are the same as those formed in the action of alkali-proteinase (trypsin) on protein, and we must in consequence ascribe to the gastric juice much greater digestive importance so far as the proteins are concerned than heretofore.

An essential difference exists, however, in the **velocity** with which the two enzymes bring about the decom**position** of protein into these simpler substances, alkali-proteina**se** acting much more rapidly than acid-proteinase. The degree of the splitting is also different in the two, trypsin causing the total cleavage of much more of the original protein than pepsin. Trypsin must, therefore, generally speaking, be considered the more powerful of the two ferments.

5. Alkali-proteinase (trypsin) is the term applied to the proteolytic enzyme found in the pancreatic juice and in a large number of tissues, not only of the human being, but other animals as well. The ferment was discovered by CLAUDE BERNARD in the middle of the last century, and has since then served as an object of study to scores of investigators. The ferment is present in the pancreatic juice of the human fœtus from before birth. Absolutely pure alkali-proteinase has never been prepared, and even relatively pure preparations are not easy to obtain. Simple extraction of the fresh, finely minced gland with a saturated sodium chloride solution gives a very active preparation. Extraction of the gland with chloroform water for several days has been recommended by SALKOWSKI. Very active and stabile alcoholic and glycerine extracts of the proteolytic enzyme can also be prepared. MAYS [1] has studied the problem very carefully, and has perhaps gotten the purest trypsin thus far obtained. For the details of his method the original must be consulted. While pancreatic juice obtained from a pancreatic fistula contains other ferments besides alkali-pro-

---

[1] MAYS: Zeitschr. für physiologische Chemie. 1903. XXXVIII. p. 428.

teinase, it contains none of the ordinary intracellular ferments present in simple extracts of the gland itself. Pancreatic juice obtained by PAWLOW'S or some similar method represents, therefore, an excellent solution with which to study the activities of the alkali-proteinase.

Alkali-proteinase will act in an alkaline, neutral, or even faintly acid medium. The medium acting most favorably has an alkaline reaction. KANITZ,[1] who has recently investigated this problem, states that the action of the enzyme is largely independent of the nature of the alkali or alkaline salt used to obtain the alkaline reaction, and is determined solely by the number of hydroxyl ions present in the solution. A 1/200 to 1/70 normal solution of the alkali in regard to the hydroxyl ions is the optimum one. An alkalinity greater than this acts deleteriously upon the enzyme. The ferment acts very well in a neutral medium, but is rapidly destroyed in the presence of an acid, even of the concentration of the hydrochloric acid of the gastric juice.

The nature of the protein upon which alkali-proteinase acts is not without influence upon the rapidity of the splitting process. Unboiled fibrin cannot well be used in making comparative tests with trypsin, as it is too rapidly digested. Boiled fibrin is digested more slowly, and this substance, or boiled white of egg, is ordinarily used in work on this ferment. Even the last-named is rapidly digested by alkali-proteinase. The protein acted upon does not first swell, as is the case with acid-proteinase, but breaks up at once into small particles which rapidly go into solution.

Alkali-proteinase is very sensitive to temperature. It acts best at about 40° C., being rapidly destroyed above this point. Below 40° C. the activity of the enzyme falls off gradually, but it is still recognizable even at 0° C. 70° C. is ordinarily given as the highest temperature at which alkali-proteinase will exhibit any proteolytic activity.

---

[1] KANITZ. Zeitschr. f. physiol. Chem., 1902, XXXVII, p. 75.

Alkali-proteinase shows, in common with other **ferments,** an increase in the velocity of reaction with an **increase in the** concentration of the enzyme. This holds **true, however,** only within certain limits of time, concentration of **both** enzyme and protein, etc. In infinite time a small **amount** of the ferment will bring about as much protein **digestion** as a larger one. In no case is *all* the protein split into the digestion products to be discussed below. Unless **the** products of digestion are removed as soon as formed **the** proteolysis is incomplete, or, as it is ordinarily put, **an** accumulation of the products of digestion interferes with the further action of the enzyme. This is probably due to the fact that the activity of the enzyme is reversible. When the products of the protein digestion are removed by dialysis as soon as formed, a small amount of the enzyme will split an indefinite amount of the protein.

When alkali-proteinase is allowed to act upon proteins, the protein molecule is broken up into a number of simpler substances. According to the generally accepted view, the decomposition of the protein is brought about by a series of successive cleavages. Shortly after the ferment has begun its work, there can be recognized the *proteoses,* and later the *peptones,* in the sense in which KÜHNE used these terms. Before the ultimate products of tryptic digestion are reached, substances which in their chemical complexity stand between them and the peptones, the *peptides* of EMIL FISCHER and ABDERHALDEN,[1] are formed. These peptides represent combinations of amino-acids, and depending upon whether two, three, four, or many molecules of the same or different amino-acids enter into the composition of the peptide, we distinguish between *di-, tri-, tetra-,* and *polypeptide* bodies. The ultimate products of tryptic digestion are *mono- and*

---

[1] EMIL FISCHER and ABDERHALDEN: Zeitschrift für physiologische Chemie, 1903, XXXIX, p. 81; ABDERHALDEN: Lehrbuch d. physiol. Chemie, Berlin, 1906, Eiweissstoffe.

diamino-acids, many in number and differing markedly from the protein itself or any of the intermediate products. In the passage from the proteins to the ultimate digestion products we find that the substances become progressively more simple. Physically we pass from those which are typical colloids, that is to say, amorphous substances with high molecular weights, practically no diffusibility, and no osmotic pressure, to crystalline bodies of a low molecular weight and of a ready diffusibility. Whereas the original substances are insoluble or only slightly soluble, the ultimate products are, generally speaking, freely soluble. Qualitatively, the digestion under the influence of alkali-proteinase does not differ materially from that under acid-proteinase; quantitatively, alkali-proteinase is the more powerful enzyme, producing the substances to be enumerated below much more rapidly and in greater quantities than when acid-proteinase is used.

In order to obtain the end products of alkali-proteinase digestion for study, it is best to allow pure pancreatic juice obtained from a PAWLOW pancreatic fistula to act upon a protein, such as fibrin. The mixture is covered with toluol or some other antiseptic which prevents the development of bacteria but does not interfere with the activity of the alkali-proteinase, and the whole is kept at a suitable temperature for varying periods of time. Many of the products of alkali-proteinase digestion can be found shortly after the ferment has been allowed to act upon the protein, but certain of the rare products cannot be found in sufficient quantities until the reaction mixture has been allowed to stand even for months. In order to get a conception of the quantitative relations between the various digestion products at any desired time, the action of the ferment can be stopped by boiling the reaction mixture. Frequently the autodigestion of the pancreas and other organs has been used for the study of the products of alkali-proteinase digestion, but in these cases it cannot be said with certainty that some of

,he products formed are due to the action of alkali-proteinase alone and not to the action of intracellular enzymes existing beside the alkali-proteinase.

The following is a list of the amino-acids that have been isolated from various proteins. In spite of the great physical differences between the proteins obtained from different sources, they are very similar in chemical constitution. When any protein is split hydrolytically, be this under the influence of alkali-proteinase or acid-proteinase, or as it is often accomplished in the laboratory, through the action of strong acids or alkalies, the same series of simple substances is always obtained which consists almost entirely of amino-acids. All proteins, moreover, yield the same amino-acids, only the proportion which these bear to each other in the different proteins is different, and sometimes one or the other of the acids may be entirely absent. In order to render what is to follow more intelligible, the next paragraph contains a list of the amino-acids which have been obtained not only through the action of alkali-proteinase, but also by other methods of hydrolysis. Most of these acids have, however, been isolated from at least certain proteins through the action of alkali-proteinase alone. It will be noticed that certain of those enumerated have already been mentioned as products of peptic digestion.

Glycocoll, alanin, aminoisovalerianic acid, leucin, isoleucin, serin, aspartic acid, glutamic acid, lysin, arginin, histidin, cystin, phenylalanin, tyrosin, prolin, oxyprolin, tryptophan.[1]

The diamino-acids in the above series, such as lysin, arginin, and histidin, are often spoken of as the *hexone bases*, and contitute with ammonia the *basic* products of the hydrolysis of proteins.

The following diagram, as arranged by ABDERHALDEN, may serve to indicate the scheme according to which the proteins are under the influence of a ferment (or other hydrolytic

---

[1] ABDERHALDEN. Lehrbuch d. physiol. Chemie. Berlin, 1906, p. 160.

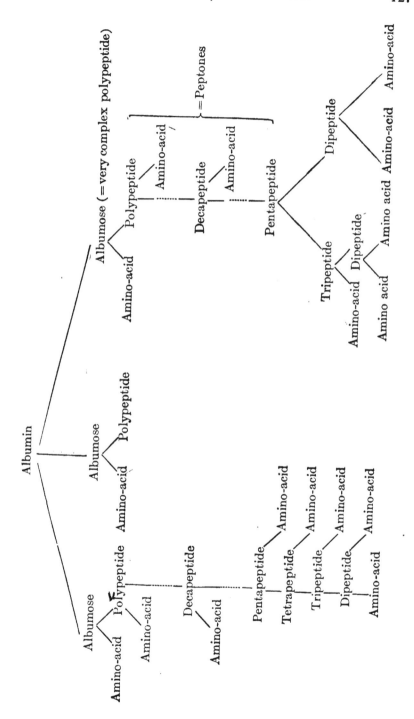

agent) broken into successively simpler compounds until the amino-acids are finally reached. From the albumins, for example, are derived first of all the albumoses, which may well be looked upon as being made up of long chains of amino-acids. These now break up into shorter chains constituting the peptones. In the analysis of albumin we are able to recognize this stage in digestion by the appearance of certain color reactions (biuret test), which it will be seen later [1] are given by certain polypeptides which have been produced synthetically. The peptones may therefore be looked upon as a mixture of polypeptides. These polypeptides now break up into still simpler chains of amino-acids passing more or less directly through the stages of hexa-, penta-, tetra-, etc., peptides until the simple mono- and diamino-acids are reached. The peptides giving a biuret test pass over without a break into those which do not give this reaction. It is clear, therefore, that little by little the name "peptone" must disappear to give way to well-defined chemical compounds. Transitional forms of all degrees of chemical complexity exist between the peptones on the one hand and the simple amino-acids on the other. The diagram does not show, of course, the way in which any protein *is* hydrolyzed, but rather one or two schemes according to which this *may* occur.[2]

---

[1] See p. 142.

[2] ABDERHALDEN: Lehrbuch d. physiol. Chemie, Berlin, 1906, p. 206.

CHAPTER VII.

THE ACTION OF THE ENZYMES FOUND IN THE HUMAN
ALIMENTARY TRACT (*Continued*).

6. The Recognition and Quantitative Estimation of the Proteinases.—The methods which have been devised for the qualitative recognition of the proteolytic ferments consist for the most part in an exposure of a readily obtainable protein, such as fibrin from blood or white of egg to an extract of the animal or vegetable organ which is being tested, and finding that this is dissolved. Care must be taken in each case, of course, to provide, through the addition of an acid or an alkali to the mixture, an acid, neutral or alkaline reaction depending upon whether acid-, ampho-, or alkali-proteinase (pepsin, papain, or trypsin) is being tested for. Instead of utilizing the disappearance of the protein as evidence of the presence of a proteolytic ferment, the appearance in the reaction mixture of certain well-established products of proteolytic activity may also be used. Since acids, for example, can by themselves bring about the destruction of a protein (but only after a long time at ordinary temperatures), the time element also plays a rôle in these qualitative tests, in that the destruction of the protein and the appearance of digestion products must occur within a relatively short time. The actual time consumed in bringing about the total solution of a given amount of protein may therefore be used as an index to the *amount* of ferment present, a larger amount of ferment, under otherwise similar conditions, bringing about a total digestion more rapidly than a smaller one.

129

· The methods which have been devised for the *quantitative* estimation of proteolytic ferments are various and still exceedingly unsatisfactory. For the most part, no absolute but only comparative estimations of the ferment content of any organ or secretion can be made. As with the other ferments, it is not possible to obtain the proteolytic ferments in anything even approximating a pure state without tremendous and incalculable loss, so that direct estimations are out of the question. We have to content ourselves therefore with comparative studies, in which we can say a definite mixture digests a chosen quantity of a protein more or less rapidly than an arbitrarily established standard. From the degree of difference we can get some idea of the relative amounts of proteolytic ferment present in the different reaction tubes, but this only within certain well-defined limits of time and limits of concentration of ferment and protein. Of the various methods and their modifications which have been devised from time to time for the quantitative estimation of proteolytic ferments, only those are mentioned of which an understanding is necessary for what is to follow in this volume.

(*a*) *Grützner's* [1] *Method.*—This is a colorimetric method in which fibrin from ox-blood is used. The fibrin obtained by whipping fresh blood is finely chopped and washed in water, after which it is stained in a carmine solution. After another thorough washing in water this carmine-stained fibrin is preserved in glycerine. When a test is to be made the colored fibrin is thoroughly washed in water and a weighed amount introduced into each of the digestion mixtures to be tested. As the fibrin is digested the reaction mixture is colored red, from the depth of which, when compared with an arbitrarily established scale, deductions can be made regarding the relative digestive power of the different mixtures.

(*b*) *Mett's Method* consists in a determination of the amount of coagulated egg albumin digested out of capillary tubes.

---

[1] GRÜTZNER: PFLÜGER'S Archiv, 1874, VIII, p. 452.

Glass tubes 1 to 2 mm. in diameter are drawn full of fresh white of egg and dipped for exactly a minute into water having a temperature of 95° C., in which the albumin coagulates. After the tube has been allowed to cool slowly it is cut into short pieces and these are dropped into the digestion mixtures to be studied. The number of millimeters of egg-white digested out of the tube is determined by the help of a low-power microscope and scale from which the digestive power of one mixture as compared with that of another may be determined. According to a principle first declared by SCHÜTZ the amount of proteolytic ferment present in the various reaction mixtures is proportional not to the number of millimeters of albumin digested out of the tubes but to their square. A digestion mixture which dissolves 2 mm. of albumin out of a tube is supposed to contain not twice as much proteolytic ferment as one which digests only 1 mm. out of a tube, but four times as much.

METT's method, though exceedingly simple, is by no means free from error.

(c) *Spriggs' Method.*—SPRIGGS [1] has devised a method for determining quantitatively the rate at which a coagulable protein is digested under the influence of different proteolytic ferments which for simplicity and accuracy is far superior to the ordinary methods employed for this purpose. SPRIGGS' method takes advantage of a physical change—a decrease in viscosity which occurs in a protein solution when this is acted upon by a proteolytic ferment. In order to measure this change in viscosity use is made of the *viscosimeter* of OSTWALD, illustrated in Fig. 19a, which allows an experimenter to determine accurately the time required for a measured amount of liquid contained in *A* to flow through the capillary tube *B* into *C*. The greater the viscosity, the greater, of course, will be the time required for a liquid to flow through the capillary tube.

---

[1] SPRIGGS: Zeitschr. f. physiol. Chem., 1902, XXXV, p. 465.

With the use of such an instrument SPRIGGS found that during digestion the viscosity of a solution of a coagulable protein gradually decreases. This decrease in viscosity corresponds with a chemical change in the protein undergoing digestion from the state in which it is coagulable by heat

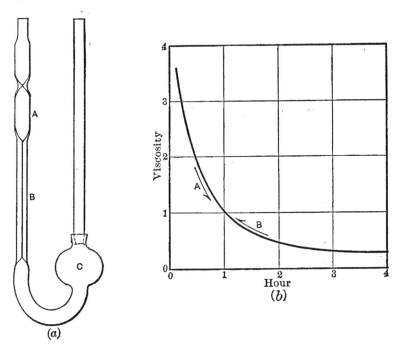

FIG. 19.

into that in which it is not coagulable by this means,—in other words (to speak not very accurately), from the albumins into the peptones.

If the change in the viscosity of the protein undergoing digestion is expressed in the form of a curve, one similar to that shown in Fig. 19 (b) is obtained.   As can readily be seen, the viscosity of the protein solution decreases at first very rapidly, later more slowly, until finally it runs almost parallel with the base-line.   When this line of unchanging viscosity is reached, the larger part of the coagulable protein has been converted into the uncoagulable.

The curve representing the decrease in the viscosity of a protein solution can be expressed mathematically when it becomes possible to calculate the relations which exist between the velocity of digestion and the amount of proteolytic ferment present.

✗ **7. Antiproteinase** (antipepsin and antitrypsin).—Under the heading antiproteinase we understand a substance discovered by WEINLAND,[1] which has the power of markedly inhibiting by its presence the action of the proteolytic enzymes. Antiproteinase is therefore one of the so-called *antiferments.* Antiproteinase can be obtained from a number of sources— the mucous membrane of the stomach and intestines, but, best of all, from various intestinal worms, especially ascaris. A description of the method by which antiproteinase can be obtained from ascaris will suffice to indicate the method by which, in general, this substance can be obtained from any of the tissues or fluids in which it is present (see below).

A very active though impure preparation of antiproteinase can be obtained by simply grinding up a number of the intestinal worms with quartz sand. The addition of this ground-up mass to an acid-proteinase or alkali-proteinase solution markedly inhibits the activity of these ferments. A better preparation of the antiproteinase can be obtained by subjecting the ascaris paste to great pressure and collecting and filtering the fluid which is squeezed out. The antiferment contained in this very active juice can be further purified by adding alcohol to it, whereby the antiproteinase is precipitated as soon as the concentration of the alcohol in the mixture exceeds 85 percent. The substances (impurities) which are precipitated before the concentration of the alcohol reaches 60 percent may be filtered off without danger of losing the antiproteinase. The precipitate of antiproteinase brought down by the 85 percent alcohol settles to the bottom

---

[1] WEINLAND: Zeitschr. f. Biologie, 1902, XLIV, p. 1; ibid., 1902, XLIV, p. 45.

and may at the end of twenty-four hours be filtered off. After washing in 96 percent, then 100 percent alcohol, and finally with ether, the precipitate is dried over sulphuric acid.

The antiproteinase obtained in this way is a somewhat sticky powder which is readily soluble in water, and which is still contaminated with some impurities. The isolation of the antiferment is accompanied by a falling off in its activity. The activity of the alcohol-precipitated antiproteinase when redissolved in water is less than half that of the original juice obtained from extraction of the ground-up worms.

The following experiments carried out with .fibrin obtained from pig's blood, which is exceedingly sensitive toward proteolytic ferments, may serve to show how markedly antiproteinase inhibits the action of alkali- and acid-proteinase.

Exp. A.   Two tubes each containing 8 c.c. water, 0.04 gm. sodium carbonate, 0.015 gm. alkali proteinase, and several pieces of fibrin, but the one in addition 7 c.c. of ascaris extract, the other an equal amount of water, are both put into the incubator at 37° C.   At the end of *two hours* the fibrin in the tube containing no ascaris extract (antiproteinase) has entirely disappeared.   The fibrin in the tube containing the antiproteinase is still entirely unchanged after *six days*, and even on the eleventh day of the experiment one-third of the fibrin is still undigested.

Exp. B.   Two tubes each containing 7 c.c. water, 0.23 gm. hydrochloric acid, 0.015 gm. acid-proteinase, and several pieces of fibrin, but the one 8 c.c. of ascaris extract, the other an equal amount of water, are put into the incubator at 37° C.   The fibrin in the tube containing no antiproteinase is dissolved at the end of *one hour*.   The fibrin in the tube containing the ascaris extract is unchanged on the *fourth day*, and on the *ninth day* only a trace of the fibrin seems to have been dissolved.

Antiproteinase is comparatively sensitive toward heat. Boiling for one and a half minutes suffices to entirely do away with the protective properties of an ascaris extract.   Heating for ten minutes to 60° C. does not injure the antiproteinase, but heating for the same length of time to 80° C. reduces its activity markedly. If the ascaris extract is kept at 95° C. for ten minutes, it has lost its protective power altogether.

At ordinary room temperature ascaris extract keeps very well when 1 to 2 percent of sodium fluoride have been added to it to prevent the development of bacteria. Under these conditions WEINLAND has kept an extract for eight months without apparent loss in the power of the juice to inhibit the action of proteolytic enzymes.

We have in the foregoing paragraphs spoken of antiproteinase as though it were a single substance. It is possible that there exist several antiproteinases, though this question is still unsettled. It may be that there exists an antialkaliproteinase (antitrypsin) and an antiacid-proteinase (antipepsin), but further experiments must be made to settle this point definitely.

8. Why the Alimentary Tract Does Not Digest Itself.—WEINLAND's discovery of antiproteinase (antipepsin and antitrypsin) is of fundamental physiological importance, giving us, as it does, a partial explanation of the immunity which "living" tissues possess towards the proteolytic enzymes. It is a well-known fact, for example, that the intestinal worms—nematodes, trematodes, cestodes, etc.—are not acted upon by the secretions of the stomach, pancreas, and small intestine, which so readily and rapidly bring about the digestion of the ordinary proteins that enter the alimentary tract as food. In the same way we know that cysticerci, the larvæ of tapeworms, must pass through the stomach in order to get into the intestine, where they develop into the adult animals. In this passage through the stomach they are subjected to the action of the gastric juice; this digests the sac in which the larva is contained, as well as the body of the larva, the head and neck only being able to pass on undigested into the intestine. From these fragments the adult develops, and this in spite of the fact that the growing animal is daily bathed in streams of intestinal contents which are charged with most active proteolytic enzymes. We know that a tapeworm may exist for months, even years, in the intestine of a man or other animal.

Similarly mysterious has appeared to us the immunity which both the stomach and small intestine possess against their own secretions. The stomach is not digested by the gastric juice, nor the duodenum when the pylorus opens. Neither is any portion of the small or large intestine, under normal circumstances, acted upon by the pancreatic juice which passes through it.

The hypotheses which have been proposed from time to time to explain this immunity of intestinal parasites and alimentary tract have been many and, for the most part, of a vitalistic nature. CLAUDE BERNARD believed that the epithelial covering of the intestinal tract protected the underlying tissues from digestion, but this idea, besides explaining nothing, stands in contradiction to the facts of pathology, which show that the absence of epithelial covering (for instance, in ulcers of the gastro-intestinal tract from any cause whatsoever) is by no means always, in fact only at times, accompanied by a loss of the underlying muscular or other tissues. These sub-epithelial structures are therefore just as immune against digestion as the epithelium itself.

Nor can PAVY's theory, which assumes that the stomach is not digested because it is protected through the alkalinity of the blood, be looked upon as any more serviceable. The blood is, first of all, not alkaline, but neutral in reaction, and, secondly, PAVY's explanation is of no value when we deal with the intestine, the contents of which at no time possess the decided acid properties of those of the stomach and are at times perhaps even alkaline. The remaining theories which have been proposed from time to time need not be discussed, for they have almost without exception covered up the problem at hand by attributing to the intact mucosa "living" properties which we could never hope to find in "dead" matter. As is well known, the "dead" mucosa of the stomach undergoes partial digestion, as seen in

the post-mortem excoriations of the gastric mucous mem·
brane.

WEINLAND'S experiments give us a ready explanation of the
long-well-established facts which have been outlined. The
mucous membrane of the stomach and intestine contains
antiproteinase; so also do the adult intestinal worms. In
the case of the cysticerci only the head and neck of the
larvæ contain antiproteinase, in consequence of which the
remaining portions of the young parasites are digested. In
the same way we may assume that the greater ease with
which cooked meats are digested than raw, is due, in part at
least, to the fact that in them the antiproteinase is destroyed
by the heat used in preparing the meat.

Many facts indicate that antiproteinase is widely distributed throughout the body, being present in probably
all tissues with the exception of the purely fatty ones. It
is entirely probable that antiproteinase exists also in the
insectivorous plants, which through their proteolytic secretions are able to digest the animals they have caught without
being acted upon by these secretions themselves. That
"living" organs, such as the spleen, when introduced into the
stomach are able to withstand the action of the gastric juice
is dependent upon the presence of antiproteinase in these
organs. In the same way, extracts of liver and muscle are
able to decrease markedly the digestive activity of acid-
proteinase, which indicates that antiproteinase is present
in them. The same is true of blood-serum and red blood-
corpuscles. The discovery that antiproteinase exists in the
liquid portion of the blood seems to indicate that this ferment
must exist in every part of the body.

Without discussing the subject more fully, which would
take us too far afield, it may not be amiss to point out what
an important physiological rôle this universal distribution
of the antiproteinase must ·play, and what pathological consequences may accompany its lack or overproduction when
it is remembered that proteolytic ferments (acid- and alkali-

proteinase) are found in practically every tissue of the body. A change in the proper balance between ferments and antiferments can easily be imagined to be accompanied by profound changes in the activities of any organ.

The probable rôle which a lack of antiferment can play in the production of gastric and intestinal ulcers deserves special mention. It is readily apparent from what has gone before that if the stomach, for example, does not digest itself because the mucosa contains antiacid-proteinase, a lack of this protective principle would allow the gastric secretion to take effect upon the mucosa as well as upon any food which passes through the stomach. What holds for the stomach is true of the intestinal tract in general. In the absence of experiments regarding this point, it is useless to discuss the causes which might lead to such a lack of antiferment in different portions of the alimentary tract. A large number of toxic causes can be imagined as capable of so interfering with the normal activity of the alimentary mucous membrane as to lead to an inadequate production of antiproteinase, or none at all. That bacterial toxins and various mineral poisons are capable of producing ulcerations of the gastro-intestinal tract is a well-known fact. Whether all or some of them act as indicated could readily be determined by experiment.

Special mention must be made of the round ulcers of the stomach. In these cases the general pathological blood condition so often found in these cases might be the predisposing condition which favors the inadequate production of antiproteinase in certain regions of the stomach. Some authors have expressed the view that an occlusion of the blood-supply of circumscribed areas of mucous membrane in the stomach lies at the basis of round ulcers in this viscus. It is certainly true that an experimental occlusion of the artery supplying the mucous membrane of the stomach or intestine is almost constantly followed by an ulcer. The lack of blood-supply would in this case have to be looked upon as the cause which led to the injury of the cells of the mucous membrane, which, with their consequent inadequate supply of antiproteinase, were digested by the proteolytic ferments present in the alimentary tract.

In cases of round ulcer of the stomach yet another factor besides lack of antiproteinase plays a rôle in the production of the lesion, namely, a hyperacidity of the gastric juice. While this hyperacidity has by several authors been looked upon as being involved in the etiology of round ulcer, just how it was involved could not be said. The experiments of WEINLAND give us a clue. While antiproteinase is able to protect fibrin against digestion by a pepsin-hydrochloric acid mixture, it is able to do so only within certain limits of concentration of the hydrochloric acid. When the concentration of the hydrochloric acid exceeds a certain point the antiproteinase is only partially able to inhibit the action of the acid-proteinase, so that the fibrin slowly goes into solution. We can imagine the hyperacidity of patients affected with round ulcer to play a similar rôle, in that the excessive acid present prevents the antiproteinase from adequately protecting the stomach-wall. But the hyperacidity can by no means be considered even the chief etiological factor, for if it were, gastric ulcers should be diffuse affairs, while they are instead more or less localized and often intimately connected with local vascular disturbances.

The resistance which the submucous tissues of the alimentary tract (such as the musculature) usually offer to the digestive fluids that bathe them must also be attributed to the presence of antiproteinase in them. In harmony with this idea we must also assume that when the ulceration of the alimentary tract extends beyond the mucosa the involved tissues have first been rendered susceptible by causes which lead to the presence of an inadequate supply of the antiferment.

The parallelism which exists between the proteinases and antiproteinases on the one hand, and the toxins and antitoxins on the other, will occur to every one. The mucous membrane of the alimentary tract is, by virtue of its contained antiproteinase, "immune" against the "toxic" proteinases which daily pass over it.

A second reason why the tissues of the alimentary tract as well as the tissues of the body in general, are not acted upon by the ferments present in them may reside in the chemical constitution of the substances making up the cells themselves. As ABDERHALDEN[1] has pointed out, the different proteins contained in an organ are by no means acted upon by the proteinases with the same ease. The substances making up the connective tissues, such as elastin and spongin for example, are acted upon scarcely at all by a pepsin-hydrochloric acid mixture or trypsin. This property is connected with the chemical constitution of the substances concerned, which contain those amino-acids in largest amounts whose presence offers the greatest resistance to hydrolysis of the protein molecule. It might well be possible, therefore, that living cells are endowed with properties which enable them to so modify the proteins absorbed by them as to render them incapable of being acted upon by the ferments contained in the cells. Only a very slight change in the chemical constitution of a compound will make it impossible for a ferment to attack it, for, as is well known, the ferments are dependent in a most limited way upon the stereochemical construction of the molecules upon which they are capable of acting.

9. **On the Reversible Action of the Proteinases.**—The question of the means by which the simple digestion-products of the proteins are again built up into the complicated albumins, globulins, etc., which we find in the organism has within the last few years been much debated. Of first interest in this connection are the researches of EMIL FISCHER,[2]

---

[1] ABDERHALDEN: Lehrbuch d. physiol. Chemie, Berlin, 1906, p. 510.

[2] EMIL FISCHER: Berichte d. d. chem. Gesellsch., 1906, XXIX, p. 530, where references to the individual papers of the author will be found.

and with him of EMIL ABDERHALDEN[1] and their pupils, on the chemical synthesis of proteins. Even though no more than perhaps a few of the methods employed by these workers to bring about these syntheses are of a character which we can imagine as possible within living cells, this work must for all time furnish the foundation upon which further advances in the chemistry and physiology of the proteins must build.

Proceeding on the hypothesis that the protein molecule represents a long series of amino-acids chemically joined to each other in various combinations, EMIL FISCHER showed that it is possible to link two or more amino-acids together and obtain a series of chemically more complex compounds known as *peptides.* If, for example, one molecule of glycocoll is combined with a second molecule of the same substance the *dipeptide* glycyl-glycin is obtained. In a similar way leucyl-leucin can be obtained through the union of two molecules of leucin, and alanyl-alanin from the union of two of alanin. All these are *dipeptides.* It is possible, however, to make three, four, five, six, etc., amino-acid molecules enter into chemical combination with each other and so obtain tri-, tetra-, penta-, hexa-, or polypeptides. A large number of such peptides have been prepared, as examples of which the following may be cited: [2]

Dipeptides: Glycyl-alanin, alanyl-glycin, alanyl-leucin, leucyl-glycin, glycyl-tyrosin, leucyl-prolin, etc.

Tripeptides: Leucyl-glycyl-glycin, leucyl-alanyl-alanin.

Tetrapeptides: Tetraglycin, dileucyl-glycyl-glycin, dialanyl-cystin.

Pentapeptides: Pentaglycin, leucyl-tetraglycin.

The means by which such syntheses are accomplished are various, and the number of combinations possible very great. Let us ask now whether they are of more than chemical in-

---

[1] See ABDERHALDEN'S excellent Lehrbuch d. physiol. Chemie, Berlin, 1906, Eiweissstoffe.

[2] ABDERHALDEN: l. c., p. 196.

terest, and whether they actually represent combinations of amino-acids such as are produced in the action of ferments or acids on proteins of various kinds.

That such is the case can be shown first of all by certain chemical tests. In the analysis of proteins we encounter a group of substances ordinarily called peptones, which are characterized by their power of giving certain reactions such as the biuret test. It is of great interest, therefore, that many of the peptides which have been produced synthetically also give the biuret test. While the dipeptide glycyl-glycin and the tripeptide triglycin give no biuret reaction, this is positive with the tetrapeptide tetraglycin. Dialanylcystin gives a biuret reaction, while the higher peptides containing seven and more amino-acids in the molecule, such as leucyl-pentaglycin, give a red biuret test which seems identical with that given by the peptones derived from silk. Phosphotungstic acid, which is abundantly used as a precipitant for the simple digestion-products of proteins, will also precipitate many of the synthetic peptides. Certain of the amino-acids which are only difficultly soluble in water yield peptides which are readily soluble. The reverse is often observed in the analysis of proteins. When, moreover, amino-acids having a sweet taste are joined together chemically they yield substances having a bitter taste. It is a well-known fact that the ordinary peptones have a bitter taste.

Certain of the peptides which have been artificially produced can be split into amino-acids by pancreatic juice in the same way as the peptides formed in an ordinary digestion mixture.[1]

With these remarks on the synthesis from amino-acids of substances which seem to agree in their chemical and biological character with the peptones we will try to see if there is any experimental evidence at hand to indicate that from substances standing close to the class of peptones others giving the reactions of the albumins may be obtained. It

---

[1] ABDERHALDEN: l. c., p. 201.

looks as though this further synthesis of albuminous substances has been accomplished and by means which we can well imagine active in the living organism.

Of first interest in this direction is the work of OKOUNEFF, who, in 1895, showed that through the action of trypsin on a concentrated solution of proteoses the latter suffers a physical change in that flakes appear in it, or in that the whole assumes a jelly-like consistency. This change in consistency (increase in viscosity) has been shown to occur not only under the influence of alkali-proteinase (trypsin), but also acid-proteinase (pepsin) and ampho-proteinase (papain). It is a change, therefore, which is brought about in a solution of proteoses by any of the ordinary so-called proteolytic ferments.

What is the character of this change in viscosity? This question has been investigated by a number of observers from both a chemical and a physical standpoint, and the results obtained by each indicate very strongly that the process of the formation of "plastein," as the above is called, represents a reversion, under the influence of the proteolytic ferments, of proteoses into more complicated proteins. From a chemical standpoint, proof in this direction has been brought especially by SAWJALOW,[1] who has succeeded in obtaining from a proteose solution upon which a proteolytic ferment had been acting for some time, and which originally had been free from this reaction, a coagulum on boiling after the addition of acetic acid. This is a reaction which, it is well known, is considered characteristic of the albumins.

The change which occurs when "plastein" is formed has been investigated from a physical standpoint by HERZOG,[2] and to him belongs the credit of recognizing in it the reversible activity of a proteolytic ferment.

---

[1] SAWJALOW: Pflüger's Archiv, 1901, LXXXV, p. 171; Centralblatt für Physiologie, 1903, XVI, p. 625.

[2] HERZOG: Zeitschrift für physiologische Chemie, 1903, XXXIX, p. 305.

As proof that the formation of "plastein" indicates that a proteolytic ferment has synthesized from the proteose solution a substance or substances which lie nearer the original protein than the proteoses, the following may be said:

We became familiar on p. 131 with SPRIGGS' observation that a protein solution undergoing digestion under the influence of acid-proteinase (pepsin) and hydrochloric acid suffers a *decrease* in viscosity, and that this decrease occurs rapidly at first and then more slowly. When a solution of proteoses and peptones (ordinary commercial "peptone") has added to it an artificial gastric juice the mixture is found to *increase in viscosity*, at first only slowly and then very rapidly. *We have, in other words, exactly the reverse of what occurred before.* This seems to justify the conclusion that acid-proteinase (pepsin) is able not only to catalyze the analysis of a protein but also to catalyze its synthesis from the products of digestion. What has been said holds not only for acid-proteinase (pepsin) but also for alkali- and ampho-proteinase (trypsin and papain).

If the above is true and the formation of "plastein" is really to be looked upon as representing a synthesis of protein which occurs under the influence of a proteolytic ferment, then we should expect that the same external conditions which favor or hinder the activity of a proteolytic ferment in hastening the analysis of a protein should favor or hinder in the same way the activity of this ferment when it is hastening the synthesis of a protein. Thanks to the researches of WEINLAND,[1] we are familiar with a substance, antiproteinase (antitrypsin, antipepsin), which retards (practically prevents) most markedly the activities of the proteolytic ferments. The addition of a small amount of anti-proteinase, obtained, for example, from the body of the intestinal worm ascaris, to a digestion mixture of protein and proteolytic ferment practically prevents the latter from act-

---

[1] See p. 133.

ing upon the protein. *In the same way ascaris extract prevents the synthesis of protein (formation of "plastein")* when it is added in a corresponding amount to a reaction mixture consisting of proteoses and a proteolytic ferment.

The constant association of a milk-curdling ferment with a proteolytic ferment—the association of caseinase with acid-, alkali-, and ampho-proteinase wherever these ferments have been isolated—has given rise to the idea to which attention has already been called elsewhere, that these two ferments are united in a giant molecule, but that different portions of this molecule have different activities. It is of interest, therefore, to add in this connection that antiproteinase, which reduces so markedly the proteolytic activity of the giant molecule, does not affect its milk-curdling power.

CHAPTER VIII.

THE ACTION OF THE ENZYMES FOUND IN THE HUMAN ALIMENTARY TRACT (*Concluded*).

10. Protease (erepsin).—The observation of SALVIOLI, HOF-MEISTER, NEUMEISTER, and others that peptone solutions when brought in contact with pieces of still living intestinal mucous membrane no longer give a biuret reaction after the lapse of some time, that is, disappear, has usually been interpreted as evidence indicating that the small intestine has the power of synthesizing protein from the products of protein digestion. It has, in other words, been generally believed that the peptones which are formed in the course of ordinary digestion are built up again into more complex bodies—those giving no biuret reaction—in their passage through the intestinal wall. Within recent years COHNHEIM [1] has repeated some of these older experiments, but in attempting to find corroborative evidence for the ordinary explanation by the discovery of a larger amount of coagulable protein in the intestinal wall or its surrounding liquids after the biuret reaction had disappeared from the peptone solution than before, his endeavors proved unsuccessful. He found instead that *in place of an increase in the amount of coagulable protein he really got an increase in the amount of crystalline digestion-products* as the biuret reaction disappeared. The peptones which are formed in the course of ordinary digestion when in contact with the living intestinal mucosa there-

---

[1] COHNHEIM: Zeitschr. f. physiol. Chemie, 1901, XXXIII, p. 451; 1902, XXXV, p. 134.

fore disappear, not because they are synthesized into more complex compounds, but because they are broken up into simpler ones. A recognition of this fact led to the discovery of protease (erepsin), to a discussion of the identification and properties of which we shall now turn.

The proof that the change from peptone to the simple crystalline compounds mentioned above occurs under the influence of a ferment contained in the wall of the small intestine can be shown very well by introducing into each of two tubes a solution of peptones and several pieces of well-washed intestinal mucous membrane from a freshly killed dog or cat. One of the two tubes is then boiled, after which both are set aside in an incubator at 39° C. for a number of hours. While in the boiled tube the biuret reaction persists even if we wait for days or weeks, it is found to become fainter and fainter in the unboiled tube, until in the course of perhaps an hour or two—depending upon the amount of peptone and intestinal mucous membrane originally present—the biuret reaction disappears entirely. Hand in hand with this disappearance of the biuret reaction, goes an increase in the amount of crystalline precipitate which may be obtained upon the addition of suitable reagents, and to a discussion of which we shall return immediately.

It is only necessary to add that this ferment, which the above simple experiment has indicated exists in the mucous membrane of the intestine, can be extracted from it. Simple extraction with an alkaline physiological salt solution of the intestinal mucous membrane, after it has been scraped from the submucosa and thoroughly triturated in a mortar with sand, suffices to yield a very active solution of the enzyme, though, of course, not a very pure one. The same difficulties which were found to exist in the preparation of the other ferments in a pure state exist here also. By fractional precipitation with ammonium sulphate much purer specimens of the ferment may be obtained, but for the details of this process the reader must consult the original.

We have now to discuss the character of the crystalline products into which protease splits peptones. If a protease solution obtained from dog's intestine, for example, is added to a solution of peptones (in Kühne's sense), and the whole is kept in an incubator at body temperature, it is found that only a slight biuret reaction is obtained on the third or fourth day from the reaction mixture which on the first day gave an intense reaction. If, now, the digestion experiment is continued a few days longer, the biuret reaction disappears entirely, indicating that the peptones have been changed into something else. This "something else" has been shown to consist of leucin, tyrosin, lysin, histidin, arginin, and ammonia, all of them substances therefore identical, qualitatively at least, with those obtained when alkali-proteinase (trypsin) is allowed to act on a protein. The question may therefore very justly be raised, Are we not perhaps really dealing with the action of alkali-proteinase in an extract of the small intestine? This question is to be answered in the negative, and for the following reasons. Alkali-proteinase, it is well known, acts upon a large number of the so-called "native" proteins—for example, fibrin, white of egg, serum albumin, etc. If proper precautions are taken to obtain a protease solution free from alkali-proteinase this property of acting on native proteins is lacking. Fibrin, for example, which is acted upon so rapidly by alkali-proteinase that it can scarcely be used in quantitative studies with this ferment, may remain in a protease solution for days, even if the fibrin has not been boiled, without showing any evidences of having been attacked.

One of the ordinary "native" proteins is, however, acted upon by protease, and that is the casein of milk. This fact is of physiological importance, as the presence of protease in the intestine renders one of the foods of infants capable of digestion even when the ordinary proteolytic ferments (acid- and alkali-proteinase) are lacking. Protease will act also upon certain of the proteoses, but its activity manifests itself par excellence upon the peptones. As the proteoses constitute, with the exception already given, the chemically most complex substances upon which COHNHEIM's ferment acts, it has been called protease.

Through the investigations of a number of authors, more especially VERNON,[1] we have become acquainted with the presence of protease in other tissues besides the small intestine. It is very probable that this fact will necessitate a revision in our knowledge of the ultimate digestion-products of alkali-proteinase, for the two ferments are frequently present in the same tissues, and no doubt some of the substances which are to-day believed to have been produced through the activity of the alkali-proteinase are really produced through the protease which is present simultaneously and which has not been considered in the analyses at present at hand. Of interest also in this connection is the fact that the "BENCE-JONES protein," which appears in the urine in certain pathological states, is not acted upon by protease. This seems to indicate that it does not belong to the proteoses (albumoses), under which heading it is usually classified.

Protease, like other ferments, is markedly affected by the character of the medium in which it is active. It acts best in an alkaline medium, although it is still able to bring about its characteristic effects in one having an acid reaction.

11. **Lipase.**—Under the term lipase (steapsin) or the lipases we understand those ferments that have the power of acting upon neutral fats of various kinds and splitting these into their constituent fatty acids and alcohols. The earlier investigators believed that the formation of an emulsion from the fat represented one of the characteristic properties of lipase. This is not true, however. The distinguishing property of lipase resides in its power of bringing about not a physical but a chemical change in the fat. While we

---

[1] VERNON: Journal of Physiology, 1904, XXXII, p. 33.

are to-day acquainted with the presence of lipase in a large variety of plant-cells, it is interesting that the discovery of lipase was originally made in experiments on the physiology of the pancreas. This organ contains large quantities of the ferment, but it is well to remember that lipase is one of the most widely distributed enzymes in the animal organism, occurring in practically every organ of the mammal. The liver, the stomach, the small intestine, the kidneys, the subcutaneous tissues, the mammary glands, the blood, and the lymph, all contain lipase.

The fats of greatest physiological importance are all of them esters of the triatomic alcohol glycerine with palmitic, stearic, or oleic acids. These fats (known as palmitin, stearin, and olein) take up water under the influence of lipase and split into glycerine and fatty acid. A quantitative determination of the activity of a lipase may therefore be made by ascertaining the amount of acid that is formed. Qualitatively the formation of the acid can be readily demonstrated by the addition of an indicator to a mixture of fat and lipase.

In laboratory studies with lipase, palmitin, stearin, and olein are comparatively little used. Other esters are acted upon more rapidly by the ferment, so that this property, in addition to their ready solubility in water, renders them more suitable for study. In this way monobutyrin, ethyl butyrate, etc., have come to be extensively used.

Pure preparations of lipase have never been obtained. The best rapidly lose in strength in the course of a few days, even when most carefully protected. Ordinarily, simple aqueous or NaCl-solution extracts are made of entirely fresh organs which are minced and ground up in a mortar with quartz sand. These extracts are then used for experimental purposes. Glycerine extracts have also been prepared.

Lipase is, in the presence of water, exceedingly sensitive to comparatively low temperatures. When dry, the lipase from the castor-oil bean will stand a temperature above 100° C.

for some time without injury, according to TAYLOR's observations. The presence of even a small amount of water
soon makes the highest temperature that this lipase can
withstand fall far below the boiling-point.

Quantitative studies show that the activity of lipase is as
dependent upon external conditions as is the activity of other
enzymes. An increase in the concentration of the lipase in
a reaction mixture is followed by an increase in the velocity
of the cleavage of the fat, but by no means in proportion to
the amount of lipase added.

The amount of fat split in a reaction mixture under the
influence of lipase increases with the time, but becomes less
in each succeeding unit of time. All the fat is never split,
even when infinite time is allowed, owing to the fact that
the activity of this ferment is reversible.

Lipase is exceedingly sensitive toward acids, losing its
activity, for example, in a very few minutes in a hydrochloric-
acid solution having the concentration of the acid in the
stomach. At the height of digestion the lipase found in the
stomach can therefore have but little or no effect upon the
fats of the food. Lipase acts best in a neutral or slightly
alkaline medium. Its activity is markedly reduced one
away with entirely through the presence of various neutral
salts.

That the activity of lipase is reversible was demonstrated
in this country by KASTLE and LOEVENHART,[1] and independently of them by HANRIOT[2] in France. KASTLE and LOEVEN-
HART were able to show that if lipase is added to ethyl butyrate
this is split into ethyl alcohol and butyric acid. The reaction is, however, incomplete. If, on the other hand, lipase
is added to a mixture of ethyl alcohol and butyric acid, ethyl
butyrate is synthesized.

This may be illustrated by the following experiment:

---

[1] KASTLE and LOEVENHART: American Chem. Jour., 1900, XXIV,
p. 491.

[2] HANRIOT: Compt. rend. Soc. biol., 1901, LIII, p. 70.

1000 c.c. of an extract of ground pancreas were mixed with 1900 c.c. of a 1/10 normal butyric-acid solution and 100 c.c. of 95 percent alcohol. To the mixture was added some thymol to prevent the development of bacteria, and the whole was kept for 40 hours at a temperature of 23° to 27° C. A similar mixture was prepared as a control, only the pancreatic extract was boiled before being mixed with the butyric acid and alcohol. At the conclusion of the experiment 25 c.c. were distilled over from each of the flasks. Had any ethyl butyrate been synthesized from the butyric acid and alcohol this distillate ought to contain it. It was found that the distillate from the first flask, containing the unboiled pancreatic extract, smelled strongly of ethyl butyrate, and after being further purified rendered water milky when poured into it, owing to the formation of the only partially soluble ethyl butyrate droplets; it formed a soap (sodium butyrate) upon the addition of NaOH, and yielded butyric acid when digested with lipase. The distillate from the flask containing the boiled pancreatic extract smelled of butyric acid, gave no turbidity when poured into water, and was unchanged through the addition of lipase. A synthesis of ethyl butyrate from butyric acid and alcohol under the influence of a substance contained in the pancreas and destroyed by heating—in other words, the synthesis of an ester under the influence of lipase—seems proved by this experiment.

Just as in the analysis of an ester (or fat) the reaction is incomplete, so, too, is the synthesis. Before all the butyric acid and alcohol have been built up into ethyl butyrate, the reaction comes to a standstill (practically speaking). We are in this case also dealing with a reversible reaction catalyzed by a ferment. The lipase acts just as the maltase did in the case of maltose and glucose—it only hastens the establishment of an equilibrium between the fat (ethyl butyrate in this case), on the one hand, and fatty acid and alcohol, on the other. Whether lipase will have an analytic or a synthetic action upon any mixture of fat, fatty acid, and alcohol is dependent solely upon the relative amounts of these substances present. Speaking broadly, the lipase will analyze fat whenever this substance is present in excess, while it will synthesize it if the fatty acid and alcohol are in excess, and it will do one or the other of these until equilibrium has been established.

HANRIOT worked not with ethyl buytrate but with monobutyrin, which under the influence of lipase is split into glycerine and butyric acid. When, under proper external conditions, lipase is added to a mixture of glycerine and butyric acid, monobutyrin is produced, as HANRIOT was able to show by isolation of that compound.

The synthesis of chemically much more complicated esters than ethyl butyrate can, however, be brought about through lipase. With the lipase obtained from the castor-bean TAYLOR[1] succeeded in synthesizing triolein from a mixture of oleic acid and glycerine. Since this same ferment is unable to split triolein and other fats completely, the reaction coming to a standstill when some 75 to 90 percent of the original fat has been broken up into glycerine and oleic acid, the conclusion seems to be justified that the lipase obtained from this source is also able to catalyze both the analysis and the synthesis of fats in a way similar to that described in KASTLE and LOEVENHART's experiments.

12. **Sucrase** (invertase, invertin) is a ferment which is characterized by its power of splitting the disaccharide, cane-sugar (sucrose) into a mixture of two monosaccharides. The monosaccharides produced are known as "invert-sugar," and consist of a mixture of equal parts of dextrose and lævulose. The change which cane-sugar undergoes under the influence of sucrase is expressed by the following formula:

$$C_{12}H_{22}O_{11} + H_2O = C_6H_{12}O_6 + C_6H_{12}O_6.$$

| Sucrose | Water | Dextrose | Lævulose |

[1] TAYLOR: University of California Publications, Pathology, 1904, I, p. 33.

Sucrase is found widely distributed throughout the vegetable kingdom. In animal physiology it finds its importance as a constituent of the intestinal juice. Its presence has also been claimed in the secretions of the mouth and stomach. In the former of these sucrase is probably not present as a constituent of the saliva itself, but is to be looked upon rather as a contamination due to the excretion of this ferment by some of the bacteria which are constantly present in the buccal cavity. The view that bacteria are the source of the sucrase of the stomach is probably also correct. Some of the authors who have concluded that sucrase is present in the pure secretion of the stomach, simply because the gastric juice has the power of inverting cane-sugar, have overlooked the fact that the hydrochloric acid has this power by itself. In fact most of the invert-sugar which is found in the stomach after a meal of cane-sugar is attributable to the action of the hydrochloric acid. Most of the cane-sugar which enters the alimentary tract seems to remain unaltered, however, until it comes in contact with the juices of the small intestine.

For study sucrase is usually obtained from yeast, a culture of aspergillus, or some other vegetable cell. The method of DUCLAUX, which probably yields the most active preparations of the enzyme, consists in growing aspergillus upon a nutrient fluid for several days, and later substituting a solution of cane-sugar for the nutrient fluid. After growing upon this for about three more days, the mould gives off its sucrase to the solution. The whole is then filtered, and the filtrate vhich represents a solution of sucrase is used for the purposes lesired. It is also possible to obtain sucrase in a dry but less active state by the method of DENATHE. Yeast, after being kept under absolute alcohol for a time, is dried, pulverized, and extracted with water. After filtering, ether is added to the filtrate and the whole shaken. A gummy mass separates, which is dissolved in distilled water, and this solution is added drop by drop to absolute alcohol. The powdered precipitate obtained in this way is separated by filtration and dried. The resulting white powder keeps for a long time, and when dissolved in water shows great diastatic activity.[1]

The inversion of cane-sugar as catalyzed by sucrase is markedly influenced by time, concentration of the ferment, temperature, and other external conditions. In infinite time a small amount of the ferment brings about as much change as a larger amount. If the products of the reaction (dextrose and lævulose) are not removed, the reaction is incomplete, and when the catalysis has come to a standstill all three substances are present in the reaction mixture. It is ordinarily said that the products of the reaction interfere with the action of the enzyme—a fact which we have learned before in the consideration of most of the other ferments. The catalysis of the cane-sugar under the influence of sucrase, and the catalysis of the same substance under the influence of acids— which bring about the same chemical change in the sucrose— differ in this regard, for no limit is reached when acids are employed, the inversion being in this case complete.

Only when the concentration of the ferment is high, the temperature low, and the experiment continued for but a short time does an almost constant relation exist between the concentration of the ferment and the amount of invert-sugar formed. This explains, for example, the results of O'SULLIVAN and TOMPSON, who believed that the catalysis of cane-sugar under the influence of sucrase was governed by the same laws as the catalysis under the influence of acids. In every experiment which is continued a sufficient length of time, the amount of invert-sugar formed in the unit of time from the sucrose in the presence of sucrase gradually diminishes. The point at which the reaction comes to a standstill varies with external conditions of temperature, concentration of the reaction mixture, etc. But only when

---

[1] See EFFRONT: Die Diastasen. Translated into German by BÜCHELER, Leipzig, 1900, I, p. 59.

the products of the inversion are removed does the catalysis go to an end.

The optimal temperature for sucrase is given by KJELDAHL as 52° C. for a preparation obtained from yeast. EFFRONT states that at 0° C. the activity of sucrase is practically nil, rises slowly up to 30° C. to increase more rapidly up to 50° C., after which it falls again. At a temperature of 65° C. the ferment is rapidly destroyed. External conditions modify very markedly the points at which the optimal temperature and the maximal temperature are attained.[1] While a dilute solution of sucrase may be kept at 52° C. for an hour without losing its fermentative activity, more concentrated solutions suffer considerably if kept at the same temperature for even shorter periods of time. At 65° C. a concentrated solution of sucrase is entirely destroyed within an hour, while a more dilute solution bears this treatment with only a partial loss of its activity. Besides these variations in optimal and maximal temperatures in the *same* preparation of sucrase under different external conditions, marked differences are also found between sucrase preparations obtained from *different* sources. The various sucrase preparations differ, for example, in their optimal and maximal temperatures, in their resistance to chemical and physical agents, etc. On this ground certain investigators have tried to establish the existence of varieties of sucrase. The same can be said of this ferment, however, which has been said of others, that when obtained from different sources the impurities accompanying the preparations are not the same. Preparations of sucrase differ, therefore, not because of specific differences in the sucrase itself, but because of specific differences in the impurities accompanying the sucrase. Certain of these act more deleteriously upon the enzyme than others, and we get in consequence the long line of differences which authors have claimed exist in the sucrases themselves.

---

[1] By maximal temperature is meant the temperature at which the fermentative reaction no longer takes place.

It has been shown by KJELDAHL that sucrase acts best in a slightly acid medium. A greater concentration of the acid, be it inorganic or organic, acts deleteriously, as does also an alkali in any concentration. In fact the activity of sucrase is markedly influenced by a concentration of alkali, the presence of which cannot even be recognized by our ordinary indicators. As the intestinal juice is neutral or if anything slightly acid, we see that sucrase acts, so far as reaction is concerned, under favorable conditions in the body.

13. **Lactase** is a ferment which has the power of acting upon the disaccharide lactose (milk-sugar) and converting it into the monosaccharides d-glucose (dextrose) and d-galactose.

$$C_{12}H_{22}O_{11} + H_2O = C_6H_{12}O_6 + C_6H_{12}O_6.$$

Lactose   Water   Dextrose   Galactose

Various yeasts usually serve as a source of the ferment which has been studied more particularly by BEYERINCK,[1] EMIL FISCHER,[2] ARMSTRONG,[3] and WEINLAND.[4] EMIL FISCHER obtained the ferment most readily by making an aqueous extract of *Saccharomyces Kefir* and precipitating the ferment with alcohol. The reversible activity of this enzyme has been demonstrated by EMIL FISCHER and ARMSTRONG, who found that milk-sugar (or at least a disaccharide closely related to it) appears in a concentrated mixture of d-glucose and d-galactose under the influence of lactase, when the whole is kept in a thermostat at 35° C.

The fact that milk-sugar is one of the commonest constituents of our food, especially in suckling animals, gives the occurrence of lactase in certain organs its physiological importance. The exceedingly contradictory statements of dif-

---

[1] BEYERINCK: Centralbl. f. Bacteriol., 1889, VI, p. 44.

[2] EMIL FISCHER: Berichte d. deut. chem. Gesellsch., 1894, XXVII, 3481.

[3] EMIL FISCHER and ARMSTRONG: Berichte d. deut. chem. Gesellsch., 1902, XXXV, p. 3144.

[4] WEINLAND: Zeitschr. f. Biol., 1899, XXXVIII, p. 606.

ferent investigators regarding the presence or absence of
lactase in different organs and secretions of the alimentary
tract have been harmonized by the careful experiments of
WEINLAND. There seems to be no question but that lactase
is present in the secretions and mucous membrane of the small
intestine of all suckling animals and in certain adult animals,
provided they are fed milk-sugar or a food containing it.
The same holds true of the pancreas. Sucklings secrete
lactase in their pancreatic juice, and the gland contains
the enzyme. Adult animals come to have lactase present in
their pancreas if they are fed milk-sugar.[1]

14. Arginase.—Under this heading KOSSEL and DAKIN[2]
have described a ferment which has the interesting property
of acting upon *arginin* and splitting this into the chemically
much simpler *ornithin* and *urea*. The ferment is widely
distributed throughout the body, though it is present in
different amounts in the various organs. The liver probably
contains the largest amount, while next in order come the
kidneys, spleen, and intestinal mucous membrane.

Arginase can be readily obtained by extracting any of the
organs named with water and dilute acetic acid, but, as with
other ferments, only a small portion of the ferment actually
present within the tissues can be gotten out. Absolutely
pure arginase has not as yet been prepared, but advantage
can be taken of the fact that it is readily precipitated by
ammonium sulphate, and alcohol and ether to free it from
many of the impurities which accompany the ordinary acetic-
acid extract.

Some idea of the readiness with which arginase acts upon
arginin can be obtained from the following experiments, in
which are indicated the amounts of this substance which
may be split in ten minutes by an extract made from 25
gms. of liver substance. In one experiment 2.7 gms. of the

---

[1] See Chapter XII.. Part 7.

[2] KOSSEL and DAKIN: Zeitschrift für physiol. Chemie, 1904, XLI, p.
341, and XLII, p. 183.

original 3.2 gms. of arginin that were added to such an extract, were split, 2.0 gms. of ornithin and 1.1 gms. of urea being recovered from the reaction mixture. In another experiment 3.3 gms. of arginin of the original 3.6 gms. added were split, 2.7 gms. ornithin and 1.2 gms. urea being recovered.

These experiments furnish an interesting example of the apparent ease with which chemical changes are brought about in a living organism, which *in vitro* cannot often be accomplished without resort to what may be termed coarse chemical procedures. The ease with which arginase splits arginin into ornithin and urea stands in sharp contrast to the difficulty with which this same chemical change is brought about under the influence of boiling mineral acids.

The experiments of KOSSEL and DAKIN give us an insight into the means by which urea is produced in the liver and various other organs of the body. Whether the urea which was found by CLAUDE BERNARD and others in the secretions of the intestine is dependent upon the presence of this enzyme in the wall of the gut, or whether we have in addition a true excretion of this substance by the intestinal canal, cannot be settled until further experiments have been made.[1]

---

[1] See Chapter XVIII, Part 1.

CHAPTER IX.

THE BACTERIA OF THE ALIMENTARY TRACT.

A DISCUSSION of the bacteria of the alimentary tract follows very naturally upon a discussion of the enzymes elaborated here, for the various bacteria contain enzymes in their bodies and secrete them. Some of these bacterial enzymes are not unlike those which are secreted by the glands of the alimentary tract. By virtue of these, the bacteria may therefore augment, in a limited way, the activities of the alimentary secretions. In large part, however, the bacterial ferments are able to bring about changes in the alimentary contents which differ totally from those brought about by the secretions of the alimentary tract proper. The substances formed in this way are usually harmless in character, though at times, through excessive production or through the formation of specific poisonous substances, they assume even a pathological importance.

The subject of the bacteria of the alimentary tract, together with a discussion of their physiological and pathological rôle, has given rise to a literature which is simply enormous. Nor do the conclusions reached by the various authors at all harmonize—a fact not strange when the complexity of the problem is recognized. For under normal and abnormal conditions practically every form of bacterium enters the intestinal tract, and when it is remembered that usually more than one kind of micro-organism is present at the same time, that the medium upon which they grow (food, etc.) is subject to the greatest variation, and that the agencies

160

active in inhibiting their growth and reproduction are practically not at all understood, it is not strange that opinions differ.

1. That bacteria *exist* throughout the alimentary tract is no longer doubted by any one. At what time do these bacteria appear? The majority of investigators are agreed that under normal circumstances the alimentary tract of new-born animals is sterile. This condition of affairs does not last long, however, for, as shown by the observations of POPOFF, SCHILD, and ESCHERICH, the fæces of young children may contain several kinds of bacteria as early as ten hours after birth, and only rarely are they absent at the end of twenty-four. BORDANO found even the colon bacillus thirteen hours after birth. These bacteria enter the intestinal tract through air, food, and bath-water by way of the mouth and rectum.[1]

2. In order to show the enormous *number* of bacteria which inhabit the intestinal tract, the figures of SUCKSDORFF[2] may be cited, who found in one milligram of normal human fæcal matter an average of 381,000 micro-organisms. Yet the figures vary considerably, even within twenty-four hours, between the extremes of 2,300,000 and 25,000 per milligram. The variation in the number is, according to SUCKSDORFF, determined chiefly by the kind of food, and but little by the total amount of fæces or the amount of water contained in them. When only sterile food is consumed, the average number of bacteria per milligram is markedly reduced. BROTZU[3] has observed that this is true for the dog also when fed only sterile food. From the total amount of fæces cast off and the number of micro-organisms contained in each milligram, the

---

[1] See, for example, J. H. F. KOHLBRUGGE: Centralbl. f. Bakt., 1901, XXX, 1te Abth., p. 17; GERHARDT: Ergebnisse der Physiologie, 1904, III 1te Abth., p. 107, where extensive references to the literature may be found.

[2] SUCKSDORFF: Arch. f. Hyg., 1886.

[3] Quoted from KOHLBRUGGE: Centralbl. f. Bakt., 1901, XXX, 1te Abth., p. 13.

total number of bacteria voided in twenty-four hours can be calculated. For this figure GILBERT and DOMINICI give as an average 12,000 to 15,000 millions; SUCKSDORFF 55,000 millions; KLEIN,[1] who probably used the best technique of the three, 8,800,000 millions.

3. The *weight* of the bacteria excreted in twenty-four hours has been calculated by KLEYN,[2] who estimates that 1.13 percent of the *dry* fæces is made up of micro-organisms and that about 293 milligrams are cast off in twenty-four hours.

4. As to the *kinds* of bacteria inhabiting the alimentary tract, the following may be said. We must distinguish first of all between those bacteria which are almost constantly present in the digestive tube and those which are present only under certain circumstances. Under the latter heading we can say that practically every known form of micro-organism has been found in some portion of the intestinal tract at some time. A large number of saprophytic bacteria are always found in the mouth. After inhaling the air of rooms in which infectious diseases have been housed, the bacteria characteristic of that disease have been recovered from the mucous membranes of the mouth, throat, and nose of those who have lived in these rooms. But even from the noses and mouths of persons following the ordinary routine of life have virulent pathogenic bacteria been cultivated. This has been done repeatedly for the ordinary pus organisms, and NOBLE W. JONES has shown that virulent tubercle bacilli are not uncommon inhabitants of these regions. For the most part these, even pathogenic, bacteria are of little or no importance. When, however, the health of the individual as a whole, or the resistance, as we are pleased to call it, of the tissues inhabited by these micro-organisms is reduced through any cause whatsoever, they are able to produce their characteristic pathological effects.

The œsophagus harbors practically the same bacteria that

---

[1] KLEIN: Centralbl. f. Bakt., 1899, XXV, p. 278.
[2] KLEYN: Cited from KOHLBRUGGE, l. c.

are found in the mouth, as the inhabitants of this region are simply carried through this tube by the swallowed saliva and food. In the stomach the number of bacteria is greatly re-duced, a large number of them, as will be shown later, being destroyed by the gastric juice. Yet by no means *all* of them are destroyed, the spore-bearing varieties being especially successful in withstanding the action of the gastric juice. Yet the other varieties—such as the ordinary pus bacteria—can also traverse the stomach without losing their pathogenic properties. It is certain that some at least of the majority of the bacteria found in the mouth can pass the stomach and be carried through the entire intestine uninjured, for MILLER[1] was able to isolate from the stools 12 of 25 different varieties which he had recognized in the mouth.

Only few *varieties* of bacteria are found throughout the small intestine. As the cæcum is approached, however, their number increases somewhat, to approach a maximum in the ascending colon. From here to the rectum the sep-arate varieties change but little.

Though the last experiments have not yet been made in this field it seems fairly certain that *most* of the strict anae-robes are never found anywhere in the alimentary tract after the stomach is passed. Of the varieties of bacteria which are found we will not mention those which are recognized as the cause of certain primary diseases, nor those which *may* be found, but only those which, generally speaking, are *almost always* present. Of these the following are the most im-portant.

As a constant inhabitant of the upper portions of the small intestine of sucklings ESCHERICH has found the *Bacterium lactis aerogenes,* which is able to exist here because of its power of splitting up the milk-sugar of the food, from which it obtains its necessary supply of oxygen. Corresponding with the ever-diminishing amount of milk-sugar in the intestine from

---

[1] MILLER: Deutsche med. Wochenschr., 1885, p. 843.

above downwards it is found that the number of bacteria of this variety also decreases. As another constant inhabitant of the alimentary tract beyond the stomach we have the *Bacillus coli communis*, which according to ESCHERICH first appears in small numbers high up in the small intestine. From here the number of bacteria of this variety increases progressively from above downwards. KOHLBRUGGE questions the presence of the Bacillus coli communis in any portion of the small intestine except the region near the ileocæcal valve. The cæcum is considered by the last-named author as the distributing depot of the colon bacillus, from which it is supposed to be carried outward toward the rectum. By the term colon bacillus are here understood all the different varieties of this micro-organism which have been described.

The two above-mentioned bacteria were for a long time held to be the only *constant* inhabitants of the intestinal canal. Since ESCHERICH's early work we have become acquainted, however, with a number of others which seem to be constant inhabitants of the large bowel, and the cause here of certain of the putrefactive changes which occur in this region of the intestinal tract and not above it. Under this heading belong the *Bacillus putrificus*, certain members of the *Proteus* group, and bacteria very similar to the *Bacillus subtilis*. The first of these is an anaerobe and the cause of the bacterial decomposition of the proteins of the food. The products of this decomposition give the fæces their odor. The absence, as a rule, of the Bacillus putrificus from the intestine above the ileocæcal valve in the adult explains why the contents of the small intestine are without odor. The fæces of children are also free from odor because their intestinal tracts nowhere harbor this bacillus. Mention must also be made of the *Bacillus acidophilus*, an acid-producing organism which, like the colon bacillus, may at times be pathogenic, at times harmless.[1]

---

[1] ESCHERICH, KOHLBRUGGE, etc.: Centralbl. f. Bakt., etc., 1901, XXX, 1te Abth., p. 73.

As more or less constant inhabitants of various portions of the intestinal tract have also been described certain spirilli and blastomyces (yeasts) as also moulds and streptobacilli. The majority of these are harmless saprophytes. It must also be mentioned in this connection that it is probable that the fæces contain bacteria which it has not as yet been possible to cultivate on artificial media.

5. The following may be said regarding the *distribution* of the bacteria in the alimentary tract. The mouth has an exceedingly plentiful bacterial flora which through the agency of the food and swallowed saliva is carried down the œsophagus. In the stomach the bacteria become greatly reduced in number. This is dependent upon the fact that the secretions of the gastric mucosa destroy or at least inhibit the growth of micro-organisms in this organ. Which constituent of the gastric juice it is that brings about this sterilization in the stomach is not entirely decided, but it seems to be the hydrochloric acid. The pepsin by itself has no apparent action upon the bacteria and simple hydro-chloric-acid solutions of the concentration found in the stomach are sufficient to bring about the same degree of bacterial destruction as is brought about by the gastric juice.

The bactericidal effect of the gastric juice is of no mean clinical importance, for, as has been shown by OPPLER, SEIFERT, MESTER, and others, gastric and intestinal fermentation is much greater in cases of anacidity of the stomach than under normal circumstances. OPPLER believes that many chronic diarrhœas are primarily dependent upon a lack of acid in the stomach, and MESTER has found that while intestinal putrefaction is not increased when putrid meat is fed to healthy dogs, this is the case when anacidity is present. To the decreased amount or total lack of hydrochloric acid is also attributable the presence of butyric and other fatty acids in the stomach contents of patients suffering from gastric carcinoma and other diseases of the stomach. These acids are produced in part through the decomposition of the fatty constituents of the food by certain bacteria which are killed when the secretions from the gastric mucous membrane are normal, in part through the action of the lipase normally present in the gastric secretion, which in the absence of hydrochloric acid can exhibit its characteristic activity.

Throughout the duodenum, jejunum, and ileum bacteria are exceedingly scarce, dependent also it seems upon a deleterious action of the secretions of these portions of the alimentary tract upon the bacteria. It is entirely probable that the scant bacterial development found in these localities is determined chiefly by the fact that the alimentary contents are neutral or faintly acid in reaction. The presence of even traces of free acids inhibits the growth of most bacteria very markedly. It must be remembered also that the food passes through the small intestine fairly rapidly, so that little time is allowed for bacterial growth.

The pancreatic juice does not seem to have any bactericidal action. DUCLAUX found bacteria as constant inhabitants of the pancreatic duct, where they are bathed in the secretions from the gland. They also develop readily on macerated pancreas. It is of interest after what has been said above that the pancreatic juice is alkaline in reaction.

Nor has the bile any great antiseptic action, for according to TALMA even but slightly virulent colon bacilli can produce a severe inflammation of the liver when injected into the gall-bladder. Whether the pancreatic and biliary secretions together, or these in conjunction with the intestinal juice, have a bactericidal action is not yet determined.

In the large intestine, beginning with the cæcum, the number of bacteria increases enormously to reach, by the time the rectum is reached, the gigantic proportions which have been discussed above. The cause of the prolific development of micro-organisms in the large intestine is by no means clear. The fact that the intestinal contents assume an alkaline reaction after the ileocæcal valve is passed no doubt plays an important rôle. Again, the time which the food spends in this part of the intestinal canal is far greater than that in any of the preceding portions, so that the time allowed for the multiplication of the bacteria is correspondingly greater. We must, moreover, remember the antiperistaltic movements of the transverse and ascending colon as a potent factor in carrying the bacteria from the descending arm of the large bowel into the transverse and ascending portions of the colon. Bacteria are constantly entering the alimentary tract by way of the rectum. Until, however, we are better acquainted with the physiology of the development and growth of bacteria, we can expect to make but little progress in this as well as in other branches of bacteriology. Systematic studies of the influence of external conditions upon bacteria, such, for example, as WILLIAM B. WHERRY[1] has recently begun on cholera and which promise so much in the advance of our knowledge of mycotic diseases, are still very rare.

The question naturally suggests itself as to how the bacteria get from the mouth into the large bowel if the stomach through which all the food passes has such well-marked bactericidal action. We answer this question at present by saying that the bacteria are either not all killed, that they pass through in the form of spores, or finally, that they are not affected by the gastric juice because they are enclosed in particles of food. But even the practically empty intestinal tract of starvation is not entirely free from bacteria, from which the conclusion has been drawn that the intestine has a flora of its own. Great care must, however, be exercised in accepting this view, as it has not yet been proved that the various bacteria found in starvation are not such as are capable of living on the secretions of the intestinal mucosa alone. As these secretions are different in different portions of the intestine it is not

---

[1]WHERRY: Journal of Infectious Diseases, 1905, II, p. 309.

strange that larger numbers of micro-organisms should also live in some portions of the gut than in others. The existence of a greater variety of bacteria in certain portions of the alimentary tract would also increase the chances for a symbiotic development here. At any rate in the absence of systematic observations on the effect of external conditions upon bacteria we must be exceedingly careful in accepting any explanation of their distribution throughout the alimentary tract which is based upon purely vitalistic conceptions.

6. Much has been and more will be written concerning the influence of the ordinary bacterial inhabitants upon the general health of the host. The majority of investigators seem agreed that under ordinary circumstances the bacteria of the alimentary tract (even the pathogenic which are found here) do not penetrate the intestinal mucosa. Under various as yet entirely unknown conditions, however, these micro-organisms may pass readily through the wall of the gut. This is very generally the case shortly after death, and during life is perhaps the cause of certain local or general infections which start from the intestinal tract. Why the intestine should under certain circumstances lose its power of holding back these micro-organisms is not yet explained. It is not improbable, however, that the taking up of bacteria by the cells of the intestinal wall is determined at least in part by the same circumstances which determine the taking up of bacteria by the leucocytes, namely, alterations in surface tension. It would not be strange to find that all those conditions which determine the entrance of bacteria into the intestinal mucosa are such as alter the surface tension of the cells composing it.

Even if under ordinary circumstances bacteria do not pass through the walls of the alimentary tract the same cannot be said of their *products*. Many of these are readily diffusible and so are absorbed. These products of bacterial activity are at times harmless, at other times intensely poisonous.

A discussion of the *products* of bacterial activity in the alimentary tract is therefore next in order. Without entering too deeply into the chemical changes which the individual varieties of bacteria are capable of producing in the intestinal contents, the alimentary flora as a whole may be looked upon as able to bring about the following well-established decompositions. The amount of such decompositions is, of course, again dependent upon a large number of external conditions, of which we need mention by way of illustration only the character and amount of food consumed, the length of time that such food remains in the intestinal tract, the reaction of the intestinal contents as determined by physiological or pathological variations in the secretions poured out upon the food, etc.

The bacteria of the alimentary tract contain in their bodies or secrete, first of all, a number of enzymes which are not unlike those which are normally poured out upon the food in its passage from mouth to anus. We need mention here only amylase, sucrase, and lactase, which like the ferments normally poured out upon the food convert starch into maltose, cane-sugar into dextrose and lævulose, and milk-sugar into dextrose and galactose. Proteolytic enzymes probably identical with acid- and alkali-proteinase (pepsin and trypsin) are also found in the alimentary bacteria. By virtue of these they not only split proteins into albumoses and peptones but even into the ultimate digestion products (mono- and diamino-acids) with which we have already become acquainted in the discussion of the proteolytic enzymes found in the alimentary secretions proper. Lipase is also found in certain of the intestinal bacteria, so that the presence of fatty acid in the intestinal contents must be attributed at least in part to the activities of these micro-organisms. At one time the presence of fatty acids in the stomach in cases of carcinoma and other diseases which may be associated with a decreased amount or entire lack of hydrochloric acid was attributed solely to the activities of lipase-containing bacteria found here. Since we have become acquainted with the presence of lipase in the normal secretions of this viscus, this idea must be modified, and the two sources of lipase be taken into consideration. Mention must finally be made of the *cytase* which certain of the intestinal bacteria contain. This ferment has the power of acting upon cellulose and converting it into dextrin and glucose.

From what has been said, therefore, it can readily be seen that certain of the bacteria of the intestinal tract may even be of service to the animal which harbors them, for they produce changes in the alimentary contents not unlike those brought about by the normal secretions, and in the last example cited above, they may even render an otherwise useless constituent of our food (the cellulose of the vegetable cells) useful to the animal organism by converting it into substances which can be taken up by the body. While these cytase-containing bacteria probably play only an insignificant rôle in the digestive processes of the carnivora, the herbivora no doubt are dependent upon these bacteria in no mean way, for even if the amounts of dextrin and glucose formed in the hours during which the bacteria are active are not very large, the dissolution of the cellulose walls of the vegetable cells liberates their more readily digestible contents and so puts these into a position to be absorbed.

So far as the amount of the just-described forms of digestion of which the various bacteria are capable is concerned, it is no doubt correct to say that as compared with the activities of the normal secretions of the alimentary tract this is, under ordinary circumstances, comparatively little.

We come now to the bacterial decompositions in the alimentary tract which are apparently of no service to the host and which are at times of a distinctly injurious character. It is these that have excited the greatest amount of medical discussion and have led to the volumes of literature on intestinal putrefaction, its qualitative and quantitative determination, its physiological and pathological importance, and the means of combating it. Of the nature of these decompositions we know as yet but little. That they are enzymatic in character is proven by numerous experiments and is generally conceded, but of the nature of the individual enzymes which are active we are almost entirely ignorant, as also of the finer chemistry of the changes which occur in the alimentary contents in their transformation into the substances which we recognize as the products of bacterial activity.

When the alimentary contents are analyzed chemically it is found that a fairly distinct division can be made between the nature of the substances found above, and those discovered below the ileocæcal valve. The bacterial decomposition products found in the stomach may be dismissed with the statement that under normal conditions none are found. This is dependent upon a number of facts: first of all the high acidity of the gastric juice which lies far above the limits inside of which most fermentative changes can take place, and secondly, the short time (at most a few hours) that the food remains in this viscus. Whatever bacterial decompositions are possible in the stomach can, therefore, never reach a high grade in the limited time allowed for such changes under normal circumstances. The physiological importance of these two circumstances manifests itself most clearly when through experiment or disease they are missing. Every day clinical observation suffices to show how the stomach contents of a patient whose gastric juice contains too little acid, or whose stomach does not empty itself except after long intervals, are teeming with bacteria and the products of their enzymatic activities. These products differ and fall into the group of those derived from the fats, carbohydrates, or proteins of the ingested food according to the conditions found in the stomach as determined by the changes in the viscus itself, the food, and the character of the bacteria present.

Our knowledge of the bacterial products present in the small intestine of the human being has been derived from analyses of intestinal contents obtained from patients suffering from intestinal fistulæ. The observers who have worked in this field all agree that the contents of the small intestine are acid in reaction, determined, however, not by free hydrochloric acid, but by fatty acids resulting in the main from the activity of the lipase produced by the pancreas and the intestinal wall upon the fats of the food, but also in part due to the action of lipase-containing bacteria. No inconsiderable amount of acid seems to be derived from the carbohydrates through the action upon them of the bacteria found in the small intestine. Among the acids formed in this way in small amounts under normal conditions, and in often enormous amounts in pathological states may be mentioned acetic, different kinds of lactic, succinic, butyric, and formic acids. The excessive formation of these acids is at once brought about when from any cause—such as insufficiency of the secretions of the alimentary tract, or insufficiency of the proper ferments in these secretions, or enjoyment of excessive amounts of carbohydrates—these are not digested and absorbed in the proper way and so become the prey of the bacteria always present here. The formation of small amounts of these organic acids need not be followed by serious consequences. Small amounts may be absorbed, oxidized in the tissues, and no evil consequences result. Others undergo further change in the alimentary tract and are converted into water, carbon dioxide, marsh-gas, and hydrogen. The organic acids produced bring about, even in small amounts, an increased peristalsis and an increased secretion of water into the intestine, so that if the condition is at all marked, frequent liquid stools, acid in reaction and ill-smelling, result. The presence of gases in the intestines also induces peristalsis. It is clear that a slight alimentary fermentation need not be an evil thing. Unquestionably the beneficent effects of the addition of vegetables to an ordinary mixed diet in bringing about a more regular discharge of the fæces is due quite as much to the fact that the celluloses, starches, and sugars contained in the vegetables are not readily attacked (because of their structure) by the alimentary enzymes and so furnish a culture medium for the intestinal flora with the production of the above-mentioned acids and gases as to any "mechanical" stimulation which a coarse food is supposed to exert.

It will not seem surprising that acute alimentary fermentations (with diarrhœa and the general symptoms of intoxication) are much more common after the enjoyment of excessive amounts of disaccharides (candy, milk-sugar) than after the use of equally large amounts of starch. Not only does the absorption of the sugar formed from starch ordinarily keep pace with its production, but the starches themselves do not furnish as favorable a culture medium for the bacteria as do cane-sugar and milk-sugar for example. These two are not only absorbed comparatively slowly, but they are readily attacked by bacteria.

Products of protein decomposition such as are generally looked upon as characteristic of the fermentative activities of bacteria are found in the small intestine either not at all or only in small quantities.

With the passage of the alimentary contents through the ileocæcal valve a complete change occurs in the type of bacterial decomposition found. Instead of the products of- the decomposition of carbohydrates we find now the products of protein decomposition. This change is determined by a number of conditions; which is primary and which is secondary cannot be easily discovered. As the alimentary contents pass along the small intestine they become progressively poorer in carbohydrates, so that by the time the ileocæcal valve is reached little or none are left to be absorbed. The reaction of the alimentary contents also changes. While above the ileocæcal valve the intestinal contents are acid, they are alkaline below it. The character of the bacterial flora as already outlined above also changes as we pass through the ileocæcal valve, but here again it is difficult to decide whether the intestinal contents determine the character of the flora or vice versa. The fact remains, however, that while in the small intestine we deal chiefly with the products of carbohydrate decomposition we have to do in the large bowel chiefly with the products of protein decomposition. By virtue of the ordinary proteolytic enzymes contained in the bacteria the undigested proteins suffer a successive cleavage into proteoses, peptones, and finally amino-acids. The decomposition does not cease here but continues in two directions. First of all the amino group may be split off so that simple organic acids are left behind. Acetic acid is in consequence formed from glycocoll, propionic acid from alanin and valerianic acid from amino-valerianic acid. Succinic acid, phenylpropionic acid, oxyphenylpropionic acid, and skatolacetic acid may also be found. Secondly, carbon dioxide may be split off from the amino-acids with the formation of such substances as cadaverin and putrescin. Through further oxidation we obtain phenol, indol, and skatol.[1] These in combination with sulphuric acid are excreted in the urine. Their quantitative estimation in the urine has in consequence been utilized as a test of the amount of putrefaction going on in the alimentary tract.

To the volatile fatty acids and aromatic compounds and to certain of the gases formed in the action of the bacteria upon the alimentary contents is due the odor of the fæces and flatus.

From what has been said in the preceding paragraphs it might be concluded that the bacteria of the alimentary canal are either always harmful or at the best of no use to the host. This is, however, not the case. It has been shown through the experiments of NUTTALL and THIERFELDER[2] that

---

[1] ABDERHALDEN· Physiol. Chemie, Berlin, 1906, p. 184.

[2] NUTTALL and THIERFELDER: Zeitschr. f. Physiol. Chemie, 1895, XXI, p. 109; ibid., 1896, XXII, p. 62.

they may even serve a good purpose, for young guinea-pigs removed from the uterus by Cæsarean section and kept in sterile vessels, given sterile air to breathe and sterile food to eat do not thrive as do animals from the same litter, born under the same circumstances but raised in ordinary air and on non-sterilized food. It was found that the latter were uniformly better nourished, heavier, and lived longer than the former. SCHOTTELIUS[1] has made similar experiments on chicks and has found that when kept under absolutely sterile conditions these fowls in twelve days increase only 25 percent in weight instead of 140 percent, as do the control animals.

---

[1] Cited from KOHLBRUGGE: Centralbl. f. Bakt., 1901, XXX, 1te Abth., p. 26.

CHAPTER X.

THE REGULATION OF SALIVARY SECRETION.

1. Salivary Fistulæ.—The salivary glands have since the earliest days of experimental physiology served as objects of investigation, partly through the fact that changes in the character of their secretions from time to time are readily apparent and partly because their superficial situation renders them easily accessible to study.

In order to determine the quantitative or qualitative changes in the saliva as it is poured out by the three pairs of parotid, submaxillary, and sublingual glands it is necessary to collect the saliva as it issues from the ducts. It is possible to accomplish this in man and various animals by inserting fine catheters into the ducts of these glands and allowing the saliva to drip into small glass tubes. Sometimes nature creates salivary fistulæ which open externally, as when salivary calculi ulcerate through the cheek or injuries of various kinds to a salivary gland or its excretory duct cause the saliva to flow out upon the skin. The study of such cases has yielded many valuable physiological data.

The older investigators. in their animal experiments, used to lay bare the different salivary glands, catheterize the ducts and collect the saliva which poured out of them. Today, however, when we have learned how much narcotics and various operative procedures interfere with the normal function of an organ we try to employ experimental procedures which do away with such disturbing influences. In a large number of experiments it is possible therefore to work on animals in

176

which through previous operation the salivary ducts have been turned outward in such a way that they pour their secretions upon the skin where they may be collected and studied. Not only do such measures bring with them a great saving of animals, but they constitute the only means by which we can obtain a true conception of the normal activity of the gland. How revolutionary are the results obtained by such experimental means which to all intents and purposes leave the laboratory animals uninjured, not only in the case of the salivary glands but all the digestive glands, will be readily apparent from the pages that follow.

We will describe first of all GLINNSKI and PAWLOW's[1] method of making a permanent salivary fistula. A dog is ordinarily used. If the parotid saliva is to be collected a small sound is introduced into STENSON's duct and under chloroform anæsthesia a circular incision is made through the mucous membrane around the orifice of the duct. The duct is carefully dissected free from its surrounding tissues and a hole is punched through the cheek by means of a sharp knife. The duct is drawn through this hole and the collar of mucous membrane is sewed into the edges of the hole. After healing is complete the parotid discharges its secretion upon the cheek. For a study of the secretions of the submaxillary and sublingual glands a similar procedure is followed, but since it is not an easy matter to separate the ducts of WHARTON and of BARTHOLIN the two are usually together sewed into the edges of the external wound. After everything is healed the secretions from the two glands can be separated by a secondary operation in which the ducts are ligatured and opened on the side nearest the gland. In order to collect the saliva which flows from the glands small flanged glass funnels are pasted over the openings by means of a laboratory paste, and from the points of these funnels small graduated tubes are suspended. A dog which has had one or two glands operated

---

[1] GLINNSKI and PAWLOW: Ergebnisse der Physiologie, 1902, I, 1te Abth., p. 252.

upon in the way indicated suffers absolutely no inconvenience, if only care be taken to have the food fed the animal, when not being used for experimental purposes, sufficiently moist.

2. **The Relation of the Nerves to the Salivary Secretions.** —Each of the three sets of salivary glands is supplied by two sets of nerves, the one being of cranial origin, the other of sympathetic. The submaxillary and sublingual glands are supplied by a branch of the facial nerve—the chorda tympani. The parotid is supplied by the auriculo-temporal branch of the trifacial nerve. The sympathetic fibres are derived in the main from the second, third, and fourth thoracic nerves, which, after passing into the sympathetic chain, ascend to the superior cervical ganglion, from which nerve fibres, chiefly of the non-medullated variety, are given off that, after following the external carotid artery, are finally distributed to the various salivary glands. Let us see now what the effect of division and electrical stimulation of these various nerves is. In spite of the many contradictory statements found in the original papers of different investigators, these all seem to agree on the following points.

If a glass catheter is introduced into the duct of a salivary gland, no or only very little secretion flows from it under ordinary circumstances. If now the *cranial nerve* supplying the gland is laid bare (for example, the chorda tympani nerve to the submaxillary gland of a dog) and divided with a snip of the scissors, no change occurs. Let, however, the peripheral end of the divided nerve be stimulated electrically, mechanically, or chemically, and the saliva is seen to move along the catheter with increased rapidity. Soon a drop falls from its end, and this is followed by another and another in rapid succession. By far the most effective form of stimulation is that with repeated induction shocks. A very weak current will bring about such an increased flow of saliva, but within certain limits the stronger the current the greater the flow of saliva. Since the stronger currents injure the nerve, the length of time during which the flow of saliva can be kept up under these circumstances is appreciably less than when a weaker current is used. In a properly arranged experiment with a moderate current a flow of saliva may be maintained an hour or more.

The submaxillary of the dog may secrete in five minutes an amount of saliva equal to its own weight, and by alternately stimulating and resting the gland 250 c.c. may be obtained in 10 to 12 hours. Toward the end of this time the rate of flow is less than at the beginning of the experiment.[1]

Some difference exists in the amount and in the character of the saliva as obtained from the different glands in different animals. In the dog stimulation of the auriculo-temporal nerve supplying the parotid by methods similar to those just described yields from one-half to two-thirds the amount of saliva obtained from the submaxillary gland. This agrees with CLAUDE BERNARD's finding that when saliva is obtained reflexly from the dog the submaxillary gland furnishes twice as much as the parotid and ten times as much as the sublingual gland.

While the composition of the saliva as obtained from the different glands when their cranial nerves are stimulated is not the same (the submaxillary saliva, for example, always contains more organic matter than either the sublingual or parotid, and the sublingual more salts than either of the other two), still it is in all cases clear, thin, and watery, and contains only from 1 to 2 percent of solid matter.

Exposure, section, and stimulation of the *sympathetic fibres* going to one of the salivary glands is followed by quite a different effect. In this case also a flow of saliva is obtained, only much less than when the corresponding cranial nerve is stimulated, and the saliva has entirely different characteristics. While histological study may show that a

---

[1] LANGLEY: Schaefer's Text-book of Physiology, 1900, Vol. I, p. 493.

secretion of saliva has occurred in a gland shortly after stimulation, as evidenced by the presence of saliva in the smaller ducts of the gland, external evidence, that is a movement of saliva in the glass catheter inserted into the duct of WHARTON, STENSON, or BARTHOLIN, is not so readily obtained. Such a movement is discoverable only after stimulation has been continued for some time and not within a few seconds, as when the cranial nerves are stimulated. LANGLEY has calculated that at the best only 1/30 to 1/60 of the quantity of saliva which would be obtained from the submaxillary gland were the chorda tympani stimulated is obtained when the sympathetic fibres of the same gland are stimulated.

In contrast to the clear watery saliva obtained through stimulation of the cranial fibres, that obtained when the sympathetic fibres of one of the salivary glands is stimulated is turbid, thick, and ropy. While in such sympathetic saliva great variations in chemical composition are also found in different animals, it can be said in general that it is much richer in organic and inorganic material than is cranial saliva, containing as it does on an average of from 3 to 4 percent solids.

Various attempts have been made to explain these differences in the chemical composition of the various salivas and in the quantities secreted. The nerves have been charged with a multiplicity of functions, but none of these explanations can as yet be regarded as satisfactory.

Stimulation of the nerves supplying a salivary gland is accompanied by a series of *accessory phenomena* which deserve notice. When induction shocks are applied to the chorda tympani nerve, for example, not only does the rate of salivary secretion increase, but the gland swells, becomes redder in color, and the efferent veins pulsate with arterial blood. The chorda tympani, in other words, carries vasomotor nerves to the gland. It was once thought that the increased secretion of saliva was due to this increased flow of blood through the gland, but this can no longer be held.

It has been found that for a short time at least, a secretion of saliva can be obtained even from the head of a decapitated animal, in other words, in the entire absence of a circulation. That such a secretion continues no longer than it does is not strange, for in the end all the constituents of a secretion, not the least important of which is the water itself, must be obtained from the blood. The increased blood-pressure, which has so often and so long been considered the determining factor in increasing the amount of the salivary as well as the amount of other secretions, can by itself be of little moment, as is shown by LUDWIG's classical observation. When the outflow of saliva from the salivary duct is prevented, the pressure in this duct may, when the chorda tympani is stimulated, rise far above that found in even the larger arteries, such as the carotid. Finally, it is possible to bring about a vascular dilation in a salivary gland and yet get no secretion. This happens when, after a dose of atropin, the chorda tympani or the auriculo-temporal nerve is stimulated. The converse of this experiment can be performed with pilocarpin. When this alkaloid is injected, a free flow of saliva is brought about, even when no change has occurred in the calibre of the blood-vessels.

Besides the change which occurs in the calibre of the blood-vessels supplying a salivary gland, there is also a change in temperature. The active gland becomes warmer. The amount of thermal change is not determined simply by the arterial blood coursing through the gland. As LUDWIG and SPIESS have shown, when one thermometer is inserted into the secretory duct of the submaxillary of a dog, a second into the efferent vein, and a third into the carotid artery, then, when the gland is excited to secretion by stimulation of the chorda tympani, the thermometers in the duct and in the vein register a higher temperature than the thermometer in the carotid. This observation is of fundamental importance, as it gives us some clue to the character of the changes which occur in salivary and other glands during activity.   Such a rise in temperature means that exothermic reactions are taking place within the gland.   These exothermic reactions may be *chemical,* and we find support for this idea in the well-known fact that an active salivary gland consumes more oxygen and gives off more carbon dioxide than a resting one.   We are dealing therefore with oxidations in the gland.   But from MATHEWS' experiments we know that lack of oxygen increases, at least within certain limits, the secretions of the salivary glands.   Under such circumstances the heat would have to come from intramolecular oxidations, in other words, from the analysis of complex compounds into simpler ones.   We meet with less difficulty if we seek the source of at least part of the heat that is set free in certain *physical* changes. Of primary importance in this connection is the well-known fact, which PAULI has studied with particular care, that colloids in absorbing water set heat free.   The swelling of mucin and other colloids found in saliva might well therefore be considered as sources of heat.   MATHEWS' observations would also find an explanation, for an accumulation of carbon dioxide or the various poisonous substances formed in living tissues in the absence of oxygen all increase the affinity of tissues for water.

We have yet to mention the *paralytic secretion* of CLAUDE BERNARD.   When both the cranial and sympathetic nerve fibres passing to a gland are cut; all secretion ceases.   After a few hours the secretion begins once more and continues for days.   Then the secretion falls off again and finally ceases. Whether the secretion is due to degeneration (and supposedly stimulation) of the nerve and ceases when this is complete, as is generally believed, or whether the atrophy which the gland gradually undergoes is the real cause of the cessation, is not yet settled.

3. The Reflex Secretion of Saliva.—Having determined the nervous paths over which impulses may reach the salivary glands and excite them to activity, we have to discover the means by which such impulses are made to traverse these nerves. It has been a long-recognized fact that the flow of saliva, which, under ordinary circumstances is not more than sufficient to keep the mouth comfortably moist, is enormously increased as soon as sweet, bitter, or dry substances are taken into the mouth, a piece of rubber is chewed, or the mind dwells for a few seconds upon the enjoyment of some food. That so many and such diverse stimuli are capable of bringing about an increased flow of saliva has naturally led to the conclusion that *any* stimulus is effective in this direction. This is, however, not the case, as can be shown very well on dogs having permanent salivary fistulæ, operated on as described in the first section of this chapter.

A mechanical stimulus is either unable to bring about a flow of saliva, or at the best the secretion of only a few drops.

If some pebbles are thrown to a dog from some distance, so that the mechanical stimulus is fairly strong, he catches them, may move them about in his mouth, even swallow a few of them, and yet no saliva flows. Or if ice-water is poured into the mouth, or some snow is thrown in, no saliva flows. But let some sand be thrown into the mouth and the saliva flows in quantities. The same is true of all substances which the dog rejects—acids, alkalies, salts, or bitters. There seems, therefore, to be a purposeful element in the secretion of the saliva, for it does not flow when substances which do not need it, such as water, are taken into the mouth, but it does as soon as substances which are to be rejected, or which need to be neutralized, diluted, or washed out of the mouth, are taken. The degree of dryness of the food determines the amount of saliva poured out upon it—the drier the food the larger the amount of saliva that is secreted. This purposeful element seems still more striking when it is found, as experiment shows, that a thin and watery saliva is always poured out upon substances which are to be rejected, while upon edible substances a saliva rich in mucin, one, therefore, which lubricates the bolus to be swallowed and facilitates its descent through the œsophagus, is s(*treted.* Still, great care must be exercised in looking upon *every* reaction of an organism, or one of its organs as eminently purposeful (as does PAWLOW[1]), and therefore for the ultimate good of the organism. We are familiar with too many illustrations of the fact that living matter may possess characteristics that are eminently dangerous to its life and well-being.

Of much interest is the connection between the physiology and what may be called the psychology of the salivary glands. We see in the facts about to be discussed a striking illustration of how the *organic* functions of an organ may be markedly influenced through the state of the mind—an illustration not without interest to the clinicist. The same series of reactions can be obtained as have been described above, when the animal's attention is simply directed to the substances in question. When the experimenter simply *pretends* to throw pebbles or snow into the mouth of the dog no secretion follows, but the saliva flows copiously if sand takes the place of the pebbles. If various kinds of food are offered the animal, saliva flows or not, just as though the animal had been really feed them. Moreover, the same qualitative variations are noted in the saliva. If the proffered food is dry, much watery saliva is secreted, while a slimy saliva less in amount is poured out when meat is offered.

Brief mention must now be made of the paths over which impulses travel to the cranial and sympathetic nerve fibres which supply the salivary glands and influence their secretions. Both these sets of nerve fibres originate from the medulla or in the pons and spinal cord just above and below this. CLAUDE BERNARD's experiments are therefore of great interest, which show that injury to certain portions of the medulla causes a copious flow of saliva. Whatever means be employed in bringing about a reflex secretion of saliva, this can only be possible through impulses passing from

---

[1] See PAWLOW: Work of the Digestive Glands. Translated by THOMPSON, London, 1902, pp. 151 and 152.

the periphery into the medulla, and from here over the various nerves to the salivary glands. Most of the peripheral stimuli which call forth a secretion of saliva originate, no doubt, in the mouth. When food is taken into the mouth, it acts upon the gustatory nerve and the glosso-pharyngeal, and these constitute the afferent nerves over which impulses reach the medulla. But many other afferent nerves exist which can bring about a flow of saliva. When the sight of food is effective in this direction, the impulses travel over the optic nerves into the large nuclei at the base of the brain, from where, apparently, connection is made with the medulla. When smell causes a reflex secretion of saliva, the impulses must reach the uncinate gyrus and from here connect with the medulla. Of interest is the great flow of saliva which accompanies feelings of nausea. In certain cases we deal apparently with a stimulation of the endings of the vagus nerve over which impulses travel into the medulla. In others the agencies effective in bringing about the feeling of nausea (as in sea-sickness) may influence the medulla, in part at least, directly. The medulla from which the nerve fibres supplying the salivary glands arise is intimately connected with the cerebral cortex. For this reason the mere thought of pleasant or unpleasant things may cause a flow of saliva. As experimental evidence of such a connection we have the well-known fact that stimulation of certain portions of the cerebral cortex causes a free flow of saliva.

We have thus far limited ourselves to a discussion of the three great pairs of salivary glands. The many small glands scattered throughout the mucous membrane lining the oral cavity give off no inconsiderable secretion as their contribution to the ordinary mixed saliva found in the mouth and poured out upon the food. We have little accurate knowledge, however, regarding the means by which the secretion from these small glands is ordinarily controlled. Apparently they are more or less independent of the central nervous system, for when the ducts of all the large

salivary glands are turned outward, sufficient saliva is still
secreted to keep the mouth moist, and according to COLIN
this may amount, even while no food is being chewed, to
from 100 to 150 c.c. an hour in the horse.

4. On the Nature of Salivary Secretion.—The impression
might readily be obtained from the preceding paragraphs·
that the secretion of saliva is primarily a nervous act. This
is, of course, not the case, and in the last analysis the secretion of saliva must be regarded as a function of the cells
themselves constituting the salivary glands, only their
activity is markedly influenced through impulses which
pass to them over the nerves connected with them. We
have already mentioned the fact that the salivary glands
will secrete (paralytic secretion) when all the nerves supplying them have been cut. It is much more reasonable
to suppose that this secretion is an expression of activity
on the part of the glands themselves than one induced
through stimuli passing into them from the cut nerves.
In a similar way modern experiments with atropin, pilocarpin, and other poisons seem to indicate that their action
is at least not limited to their effect upon the nerves, if perhaps they do not act solely upon the secreting epithelium itself.

We know as yet nothing definite regarding the nature
of the process of salivary secretion, any more than we know
regarding the forces at work in any process of secretion.
It was once believed that the saliva represented nothing
but a filtrate from the blood which was squeezed through
the gland-cells under the influence of the blood-pressure.
This idea must be given up entirely, for the chemical composition of the saliva differs from that of blood-plasma
or lymph, not only quantitatively but also qualitatively.
Certain salts are present in greater, others in less amount
than in the blood, and while some chemical constituents
found in the blood do not appear at all in the saliva, the
reverse is also true. We must therefore conclude that the
gland cells themselves have the power of forming new chem-

ical compounds from those brought them in the blood, and secondly a power of selection in that they take out of the blood and secrete into the salivary ducts only certain of the constituents of the blood-plasma. How little the blood-pressure *by itself* is of any importance in these processes of secretion is indicated not only by the facts already cited, that an increase in blood-pressure need not be followed by an increased secretion and *vice versa*, but also by LUDWIG's classical observation. If one mercury manometer is tied into WHARTON's duct, while another is connected with the carotid artery, and the chorda tympani nerve is stimulated electrically in order to bring about a secretion of saliva, the pressure registered in the salivary duct may be 100 to 200 mm. higher than that in the artery. ✝It is possible that osmotic forces brought into play through a breaking down of complex molecules into simpler ones may explain a part of the phenomena observed, but it seems much more probable that the great pressures produced through the swelling of colloids (such as mucin) in water are chiefly responsible. Not only are colloids having a great affinity for water produced in the salivary glands, but histological evidence is at hand to indicate that during secretion those portions of the gland-cells lying farthest from the nucleus suffer the greatest changes in size. Since the nucleus is intimately connected with processes of intracellular oxidation, it is conceivable that in those portions of the cell lying nearest the lumen substances are formed which particularly favor the imbibition of water by the colloids formed in the cells. In this way the cells at first increase in size, and pressures are produced which serve to squeeze certain portions of the cell contents (the saliva) into the glandular ducts. Finally, it is not impossible that substances capable of swelling may be secreted into the smaller salivary ducts and that the pressures registered in the manometer connected with the main salivary duct may be due, in part at least, to the subsequent swelling cf these substances.

CHAPTER XI.

THE REGULATION OF GASTRIC SECRETION.

1. Gastric Fistulæ.—For our earliest knowledge of the secretion of gastric juice by the stomach, and its qualitative and quantitative variations under different physiological conditions, we are indebted to the American physician BEAU-MONT. BEAUMONT made his observations upon the hunter ALEXIS ST. MARTIN, who retained, in consequence of a gun-shot wound, a permanent opening in the abdominal wall which led directly into the stomach.

An attempt to reproduce the same condition of affairs in animals led to the experiments of more modern observers, who created gastric fistulæ artificially in animals of various kinds. The results obtained, however, were by no means harmonious or satisfactory. The animals sickened and died, or at the best secreted a juice which was evidently subnormal, both in quantity and quality. In order to obtain a flow of gastric juice, BEAUMONT introduced into the stomach of his patient various kinds of food. The students of gastric physiology who immediately followed him used the same methods, and we still use them today in the clinical examination of the gastric juice. A patient is fed a specified diet and the gastric contents, which are subsequently removed through introduction of a stomach-tube, are subjected to chemical analysis. This procedure does not, however, yield a pure juice, but one mixed with food particles, saliva, etc.

The first to try to obtain pure gastric juice was KLEMEN-

SIEWICZ,[1] who in 1875 followed the principle adopted by THIRY for the intestine, and attempted the isolation of a part of the stomach into a closed pouch which opened externally. His dog lived, however, only three days. HEIDENHAIN [2] soon after repeated the operation and succeeded in keeping his animal alive. Pure gastric juice can be obtained from a dog operated upon in this way. It is only necessary to introduce food into the large stomach, when a flow of perfectly pure gastric juice will take place from the isolated cul-de-sac.

The operation of HEIDENHAIN possesses the important defect of interfering with the nerve-supply of the stomach. To overcome this objection PAWLOW and CHIGIN [3] have devised an operative procedure which will be described in some detail, as its use has done much to give us a clearer insight into the physiology of the stomach. An incision is made into the stomach, which begins in the fundus a little below the pylorus and runs longitudinally toward the cardia. This incision divides both anterior and posterior walls (*AB*, Fig. 20, I). The triangular flap thus formed is made into a cylinder, the orifice of which is sewed into the abdominal wall, while its base is still connected with the main cavity of the stomach. The cavity of the stomach and that of the pouch do not, however, communicate with each other, but are separated by a septum of mucous membrane, as indicated in Fig. 20, II. The opening in the main stomach cavity is closed by a line of sutures.

In order to obtain gastric juice from the miniature stomach (*S*, Fig. 20, II) a small India rubber or glass tube, freely perforated at its lower end, is introduced through the opening in the abdominal wall (*AA*) into the pouch. The tube remains

---

[1] KLEMENSIEWICZ: Sitzungsberichte d. Wiener Akad., 1875, Bd. LXXI.

[2] HEIDENHAIN: Pflüger's Archiv, 1878, XVIII, p. 169.

[3] PAWLOW and CHIGIN: PAWLOW's Work of the Digestive Glands. Translated by THOMPSON, London, 1902, p. 11.

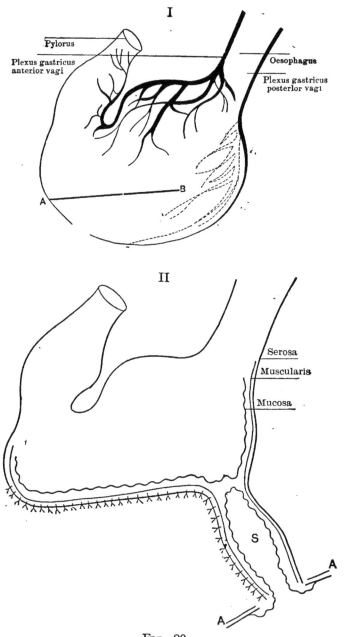

FIG. 20.
(Copied from PAWLOW: Work of the Digestive Glands. Trans. by
THOMPSON, London, 1902, p. 12.)

in the pouch of its own accord, or may be fastened there by an elastic band encircling the body of the animal. Juice is collected while the animal is lying down or is supported in an erect position in a suitable frame.

It is only necessary to add that the miniature stomach gives a true picture of the secretory activity of the large one (*V*), even though food does not at all enter the small cavity. This has been proved by a long series of carefully conducted experiments, in which the quantity and quality of the juice poured out in the large cavity has been compared with that poured out in the isolated cul-de-sac.

We have yet to describe an operative procedure which is made use of in *sham feeding*, and upon which several physiological facts concerning the stomach and its activity are based. In 1889 PAWLOW and SCHUMOW-SIMANOWSKI performed the operation of œsophagotomy on a dog already possessing a simple gastric fistula. The œsophagus is divided in the neck, and the divided ends are made to heal separately into the angles of the skin incision. This separates the stomach and mouth entirely. Dogs operated upon in this way recover and continue in excellent health for years afterward. To keep such animals alive, food must, of course, be introduced directly into the stomach.

The following experiment shows how pure gastric juice can be obtained from such an animal. If the dog be given meat to eat, the food drops out again from the lower extremity of the upper segment of the divided œsophagus. But the perfectly empty stomach begins to secrete gastric juice and this continues as long as the sham feeding is kept up. As will be shown in detail later, this is a psychic secretion of gastric juice. Suffice it for the time being to state that several hundred cubic centimeters of pure gastric juice may be obtained daily in this way without injury to the dog.

Œsophagotomy may be combined with the already described operation of PAWLOW and CHIGIN for the separation of a miniature stomach from the large. This and other operative variations in the experiments and the deductions which may be drawn from them are given below.

2. The Effect of Diet on Gastric Secretion.—The experimental methods of PAWLOW, SCHUMOW-SIMANOWSKI, and CHIGIN described above have answered for us a number of questions connected with the secretion of the gastric juice which the older methods could only hint at. Thanks to the ingenious combination of an œsophagotomy with a simple gastric fistula, and the clever surgical procedure which allows the separation of a miniature stomach opening externally from the main stomach, the secretory activity of this organ may be followed in great detail.

The stomach of the fasting animal is entirely empty. The secretion of gastric juice is dependent upon the taking of food. This can be shown very nicely in a dog possessing an isolated miniature stomach. While fasting, this is entirely empty, but within a few minutes after food is given to the animal it begins to secrete. The *quantity* of juice secreted is almost exactly proportional to the amount of food ingested. CHIGIN gives the following values to corroborate this statement:

100 gms. meat	26.0 c.c. gastric juice
200 " "	40.0 c.c. " "
400 " "	106.0 c.c. " "

When, instead of the above, various amounts of a mixed diet made up of meat 50 gms., bread 50 gms., milk 300 c.c. are given, the same fact is brought out.

The above mixture yielded 42.0 c.c. juice.
Twice the above mixture yielded 83.2 c.c. juice.

The gastric secretion is not all poured out at once upon the food, but continues as long as food remains in the stomach. The *rate* of the secretion varies, however, from hour to hour. The secretion reaches its maximum within the first hour, after which it decreases steadily in amount until at the end of a number of hours it has fallen to the zero-mark once more. The cause of this decrease is perhaps explained by the gradual decrease in the amount of food undergoing digestion in the stomach. The following experiments of CHIGIN illustrate the above:

RATE OF GASTRIC SECRETION AFTER FEEDING 100 GMS. MEAT.

Hour after feeding.	Quantity of juice in c. c.	
	Exp. *a*.	Exp. *b*.
1	11.2	12.6
2	8.2	8.0
3	4.0	2.2
4	1.9	1.1
5	0.1	a drop
Total.....	25.4	23.9·

These values are represented graphically in the following curves (Fig. 21).

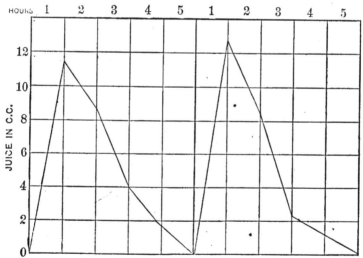

FIG. 21.—Curve of secretion of gastric juice after a meal of meat.
Two experiments.
(Copied from PAWLOW: Work of the Digestive Glands. Trans. by
THOMPSON, London, 1902, p. 23.)

PAWLOW and CHIGIN's miniature stomach allows also a study of the *qualitative* variations in the gastric juice under different conditions of diet, etc. To do this the animal is fed in the ordinary way, and the secretion which pours out of the miniature stomach is collected, measured, and used for analysis. It was pointed out above that the secretions of this small stomach are identical with those of the large.

The following table indicates how the *digestive power* of the gastric juice varies during the period of digestion, after a single feeding of 400 gms. of meat.[1] The cause of these variations is not entirely clear, in fact their very existence is questioned by some observers. The digestive power is expressed in terms of millimeters of coagulated egg-albumin digested out of capillary tubes in the unit of time. (METT's method of determining proteolysis quantitatively.)[2]

Hour after feeding.	Mm. of egg-albumin digested.	
	Exp. *a*.	Exp. *b*.
1	6.0	5.8
2	4.3	4.1
3	3.4	3.4
4	3.5	3.0
5	3.8	3.8
6	3.0	3.1
7	3.6	3.5
8	3.9	4.5

These figures are plotted in the form of curves in the following illustration (Fig. 22).

The digestive power of the gastric juice, which PAWLOW and his pupils look upon as an expression solely of the amount of *ferment* (this is certainly erroneous) contained in the juice, does not vary with the amount of juice secreted in the unit of time. A strong digestive power may be found, not only when the secretion is scanty but also when it is copious.

So far as the *inorganic* constituents of the gastric juice are concerned, PAWLOW believes that the concentration of *hydro-*

---

[1] LOBASSOFF: PAWLOW's Work of the Digestive Glands. Translated by THOMPSON, London, 1902, p. 29.

[2] See p. 130.

chloric acid never varies. *This applies, however, only to the gastric juice as it is poured out by the glands.* Once the juice has left the crypts of the gastric mucosa, its acidity may be reduced through various agencies, and it is to these that the fluctuations in the acidity of the gastric juice under various physiological and pathological conditions must be attributed. The mucus secreted by the stomach-wall has the power of neutralizing the acid of the gastric juice. For this reason

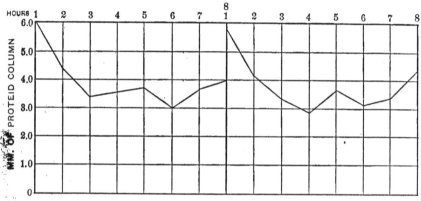

FIG. 22.—Digestive power of hourly portions of gastric juice after a meal of 400 gms. of meat.
(Copied from PAWLOW: Work of the Digestive Glands. Trans. by THOMPSON, London, 1902, p. 28.)

the first juice collected from the stomach of a fasting dog always has a lower acidity than later specimens, for the empty stomach is covered with a layer of mucus. Also, the more rapidly gastric juice is secreted the higher is its acidity, for in this way less time is allowed for the neutralization of the hydrochloric acid through the gastric mucus. Finally, the power of the saliva which enters the stomach to neutralize the acid secreted there must also be borne in mind.

The gastric juice poured out upon *different kinds of food* varies not only in quantity but also in quality. The experiments of some of the older observers lend support to this idea, but their results cannot be looked upon as entirely free from criticism. The results of PAWLOW and his

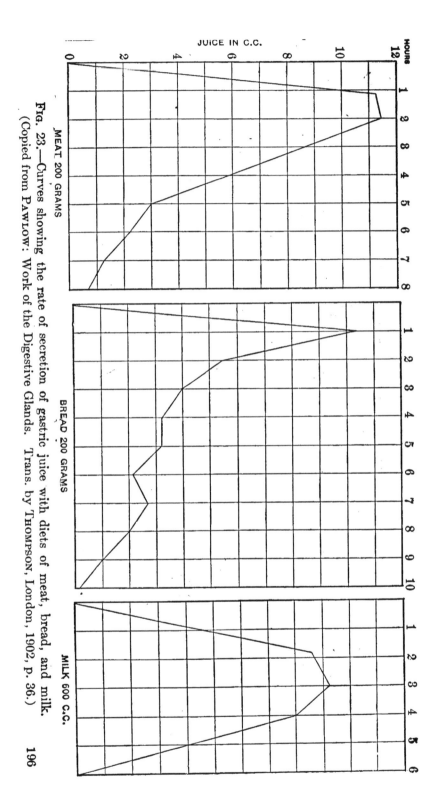

FIG. 23.—Curves showing the rate of secretion of gastric juice with diets of meat, bread, and milk. (Copied from PAWLOW: Work of the Digestive Glands. Trans. by THOMPSON, London, 1902, p. 36.)

196

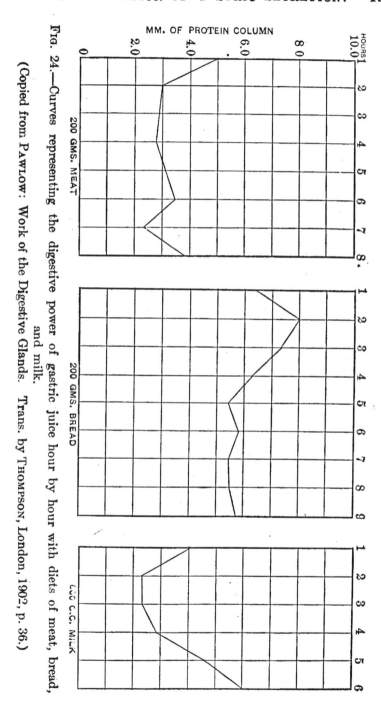

FIG. 24.—Curves representing the digestive power of gastric juice hour by hour with diets of meat, bread, and milk.

(Copied from PAWLOW: Work of the Digestive Glands. Trans. by THOMPSON, London, 1902, p. 36.)

pupils are perhaps more trustworthy. The average variation
in the ferment content (better digestive power) of the juices
poured out when equal weights of bread, meat, or milk are
fed, expressed in terms of millimeters of coagulated egg-
albumin digested out of capillary tubes in the unit of time,
is shown in the following table:

> Bread 6.64 mm.
> Meat 3.99 mm.
> Milk 3.26 mm.

Since the amount of proteolytic ferment contained in
the compared juices is proportional to the square of the
rapidity of digestion, we find upon calculation that the juice
poured out upon bread contains four times as much acid-
proteinase as that poured out upon milk and about three
times as much as that poured out upon meat, for the squares
of the figures given in the above table stand as 44:16:11.

The quantitative variation in the secretion of the gastric
juice (expressed in cubic centimeters) when different foods
are given is indicated in the following experiment of CHIGIN;
the digestive power of the juices (expressed in millimeters
of digested egg-albumin) is given in the parentheses following these values.

QUANTITY AND QUALITY OF GASTRIC JUICE SECRETED UPON FEEDING.

Hour.	200 gms. meat.	200 gms. bread.	600 c.c. milk.
1	11.2 (4.94)	10.6 (6.10)	4.0 (4.21)
2	11.3 (3.03)	5.4 (7.97)	8.6 (2.35)
3	7.6 (3.01)	4.0 (7.51)	9.2 (2.35)
4	5.1 (2.87)	3.4 (6.19)	7.7 (2.65)
5	2.8 (3.20)	3.3 (5.29)	4.0 (4.63)
6	2.2 (3.58)	2.2 (5.72)	0.5 (6.12)
7	1.2 (2.25)	2.6 (5.48)	... (.. .)
8	0.6 (3.87)	2.6 (5.50)	... (....)
9	... (....)	0.9 (5.75)	... (....)
10	... (....)	0.4 (....)	... (....)

The values are represented graphically in Figs. 23 and 24.
Each kind of food seems to bring about a definite hourly

rate of secretion and a characteristic alteration in the properties of the juice. With a meat diet the maximum rate of secretion occurs during the first or second hour, during which periods the quantity of juice secreted is approximately the same. With bread the maximum secretion occurs during the first hour, while with milk it occurs during the second or third hour. The juice has the greatest digestive power during the first hour when meat is fed, during the second and third when bread is given, and during the last hour when milk is the food furnished the dog.

As the gastric juice acts chiefly upon the protein constituent of the food, PAWLOW has made an interesting calculation of the comparative work done by the stomach in digesting the three articles of food mentioned above, taking into consideration the nitrogen content of the foods and the quality and quantity of gastric juice poured out upon them. He finds that the number of ferment units (obtained by multiplying the squares of the numbers representing the digestive strengths by the number of c.c. poured out upon the food) required for the digestion of corresponding nitrogen equivalents in the different kinds of food is as follows:

Bread 1600 units
Meat 430 "
Milk 340 "

This means that protein in the form of bread requires five times more acid-proteinase for its digestion than is poured out upon the same amount of protein in milk, and that the protein of meat requires a fourth more than its equivalent contained in milk. Now it is known that vegetable proteins are much less easily digested than those of meat, and these less than those of milk. The different kinds of proteins seem, therefore, to call forth the secretion of quantities of ferment which correspond with the differences in their digestibility.

The concentration of the *hydrochloric acid* in the gastric juice also varies with the different kinds of food. It is greatest when meat is fed (0.56 percent HCl) and lowest when bread is given (0.46 percent HCl).   A milk diet gives an intermediate figure.

3. The Relation of the Nervous System to Gastric Secretion.—The relation of the nervous system to the secretion of the gastric juice by the stomach has for many years been the subject of debate.   Against the well-known clinical fact that emotional states, injuries of various kinds, etc., profoundly alter the activity of the stomach stood the observations of a score of experimenters who were able to prove no direct connection between the central nervous system and the digestive organ.   In 1852 BIDDER and SCHMIDT published the fact that the mere sight of food will call forth a secretion of gastric juice in the dog, and in 1878 RICHET demonstrated on a patient gastrotomized for an incurable stricture of the œsophagus that a secretion of gastric juice occurred whenever the patient took certain articles of food into the mouth.   The efforts, however, to show the paths of nervous connection in these cases were singularly unsuccessful.   In recent years, however, PAWLOW and his coworkers SCHUMOW-SIMANOWSKI, JURGENS, and SSANOZKI, in a series of beautiful experiments, have been more fortunate, and, thanks to their researches, we are now familiar with the cause of failure in the older experiments, and may look upon the connection between central nervous system and stomach as experimentally proved and the nervous paths constituting this connection as fundamentally established.

The following experiment shows that the effect of feeding is transmitted to the gastric glands through nervous channels and that one of these channels is the vagus nerve.[1]   A dog, possessing an ordinary gastric fistula and œsophagotomized as described above,[2] so that the mouth is entirely cut off

---

[1] PAWLOW: Work of the Digestive Glands.   Translated by THOMPSON, London, 1902, p. 50.

[2] See p. 191.

from communication with the stomach, is used. In addition the right vagus nerve is divided below the recurrent laryngeal and cardiac branches, at the time that the gastric fistula is made. No juice flows from the gastric fistula in such a dog. If now sham feeding is indulged in, that is to say if the dog is fed food which it swallows but which never reaches the stomach, because the swallowed masses drop out of the œsophageal opening in the neck, a stream of gastric juice, which steadily increases in volume, appears at the gastric opening within five minutes after the dog is given its first food. As long as the dog is fed, which may be two or more hours, often even five or six, the gastric juice continues to pour out of the fistula. In this way several hundred cubic centimeters of juice may be collected.

How is this effect of feeding carried from the mouth to the stomach? If the food is taken away from the dog in the experiment just described, the secretion of gastric juice continues for several hours, gradually becoming less in quantity. The right vagotomy threw out of function only the pulmonary and abdominal branches on the side operated upon while the laryngeal and cardiac fibres were left intact. If now the left vagus nerve, which has been laid free in the neck by operation some three hours before the sham feeding is to be indulged in, be carefully drawn out of the wound and divided with a snip of the scissors, the pulmonary and abdominal branches of both vagi are paralyzed. The preservation of the laryngeal and cardiac fibres on the right side, however, prevents the symptoms of cardiac and laryngeal distress which follow division higher in the neck. If food be now given this dog a second time, it eats greedily as before, but, in sharp contrast to the preceding sham feeding, not a single drop of gastric juice flows from the stomach. Indeed, never again in such a double vagotomized dog, even if it lives for months afterward, as some of them do, does a sham feeding call forth a secretion of gastric juice.

When sham feeding no longer produces a secretion of gastric juice after double vagotomy, it does not mean, however, that the gastric glands have lost the power of secretion. As will be shown later, highly active juice can, under appropriate circumstances, still be obtained from this digestive organ. The above experiment only shows that certain exciting influences which normally, in the ordinary process of eating, reach the gastric mucosa by way of the vagi have been removed. As has been shown by KETSCHER, the mode of feeding and the character of the food presented to the dog during sham feeding alters markedly the character of the gastric juice obtained. If the dog is given pieces of meat at long intervals, less gastric juice and one lower in digestive power is obtained than when the dog is fed more rapidly. In a similar manner, a diet of meat, which the dog relishes highly, produces more juice and one having a higher digestive power than a meal of bread, which the dog relishes less.

The existence of nerve fibres in the vagus which influence the secretion of gastric juice was proven above by showing that after division of both vagi stimulation of the buccal cavity with food no longer excited the gastric mucosa to activity as before. It can, however, be shown by direct stimulation of the vagus in a properly performed experiment, that this nerve contains fibres which influence the activity of the stomach. This has been accomplished by PAWLOW and SCHUMOW-SIMANOWSKI. A gastrotomized and œsophagotomized dog, in which the right vagus nerve has been cut below the origin of the inferior laryngeal and cardiac fibres, is used for the experiment. Three or four days before stimulation of the vagus is to be carried out, the left vagus is carefully dissected out in the neck, a ligature passed around it but not tied, and the whole carefully preserved under the skin. On the day the vagus stimulation is to be carried out, the wound is painlessly opened and the nerve laid bare. By attending to these details, whereby appreciable pain to the animal is avoided, excitation of the vagus by induction shocks at intervals of one or two seconds invariably yields a secretion of gastric juice from the stomach. The negative or at best uncertain results of the older observers upon stimulation of the vagus are probably all to be explained by the fact that their animals were under the influence of anæsthetics or in pain, and, as will become apparent when the pancreatic secretion is described, peripheral stimuli of various kinds inhibit markedly the activity of the digestive glands. When all peripheral stimuli are prevented from inhibiting reflexly the activity of the stomach (this can be done by dividing the spinal cord just below the medulla oblongata), the vagus nerves may be laid bare in the neck and stimulated at once, when a secretion of gastric juice will be observed. In this way USCHAKOFF has succeeded in demonstrating, in an experiment performed at one sitting, the relation of the vagus nerve to the stomach.

Both forms of experiment, the "chronic" as well as the "acute," show, therefore, that the vagus nerve contains fibres which influence the secretion of the gastric glands. As will be shown later, however, a secretion of gastric juice occurs also when the vagi are cut; this indicates that the integrity of the vagus is not the only requisite for the secretory activity of the stomach. Under *normal* circumstances, however, these nerves play the already described exceedingly important rôle in the *initiation* of the gastric flow.

4. The Appetite as an Excitant of Gastric Secretion.— It was shown in the preceding paragraphs that the vagus nerve has a marked influence upon the secretion of juice by the stomach. We have now to answer the question, How is the vagus nerve normally excited? An explanation which first suggests itself is that we are dealing with a reflex excitation of the gastric mucosa, brought about through a stimulation of nerve endings in the mouth and carried from here to the vagus centre in the medulla, and from there downward to the stomach. This idea has been carefully tested experimentally by PAWLOW and found to be incorrect, as shown by the following facts.   The food upon entering the mouth is able to stimulate the buccal mucous membrane, either chemically, mechanically, or in both of these ways.   Application of such substances as acids, salts, bitters, pepper, mustard, etc., to the mucous membrane of an œsophagotomized dog never calls forth a secretion of gastric juice, however.   Not even does a meat decoction in most instances prove effective in this regard.   Nor does a combination of mechanical stimulation with the chemical, such as wiping out the mouth with an acid-soaked sponge, or giving the dog stones to swallow, work any more successfully in exciting the gastric mucous membrane.   The stones drop out of the œsophagus, and, even if this play is kept up for hours, not a drop of gastric juice flows from the gastric fistula.   As soon, however, as the old experiment of sham feeding is tried, and meat or bread is given the dog instead of stones, a free flow of gastric juice begins in five minutes and steadily increases in amount as described above.   These facts prove clearly that the nerves of the stomach are not excited reflexly through chemical or mechanical stimulation of the buccal mucous membrane. Wherein then does the sham feeding with food differ from that with stones?   In the former case the dog eagerly desires its meal, something which is lacking in the latter.   *This eager desire for food, the appetite in other words, is the excitant of the gastric flow,* and we must conclude in consequence that a psychic state rather than a reflex from the mouth acts as the normal excitant of the vagus nerve and the gastric mucosa at the beginning of a meal.

BIDDER and SCHMIDT observed years ago that merely offering a hungry dog food excited a flow of gastric juice.   PAWLOW has confirmed and elaborated this experimental finding. Actual sham feeding with food does not need to be indulged in in order to obtain a flow of gastric juice.   If only a tempting meal be prepared before a hungry dog, a flow of gastric juice, such as has been described when sham feeding is practised, begins within five minutes after the teasing is begun.   In fact it has been shown experimentally by SSANOZKI that merely tempting a dog with food often leads to a greater secretion of gastric juice in the unit of time than the actual sham feeding.

The "appetite juice" varies both in quantity and quality. The more eagerly a dog eats the greater the amount of juice and the higher the digestive power. A good appetite at the beginning of a meal is, therefore, equal to a copious secretion of strong gastric juice.

5. The Physiological Importance of the Appetite Juice.— It was shown above that the secretion of gastric juice during the first hour is practically the same for meat or bread, and that only subsequently the amount and quality of the gastric secretion varies with the nature of the ingested food. When only sham feeding with these same foods is indulged in, we find that the secretion of juice for the first hour is the same as though the dog had been actually fed. All this indicates that the first outpouring of juice into the stomach at the beginning of a meal is the consequence of the psychic excitation. The truth of this statement is still further borne out by the fact that when a food is given the dog which does not interest him to the same degree as meat or bread,—for example, milk,—this initial rise in the quantity and quality of the gastric juice does not appear. While 12.4 c.c. of gastric juice, having a digestive power of 5.43 mm., are poured out upon meat, and 13.4 c.c., having a digestive power of 5.37 mm., are poured out upon bread during the first hour after feeding, only 4.2 c.c. of a digestive power of 3.57 mm. are poured out upon milk (CHIGIN). That the mere act of taking food increases both the quantity and quality of the gastric juice is still further supported by the observation of KOTLJAR and LOBASSOFF, who found that when an ordinary meal is divided into several portions, and these are fed at intervals of three hours, an increase in the quantity and quality of the juice immediately follows each installment.

The rôle of the appetite juice can be still more clearly

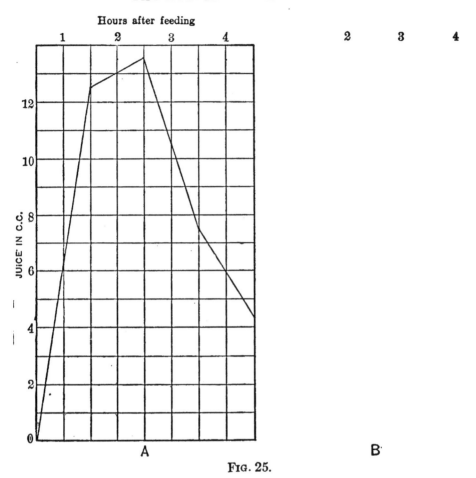

FIG. 25.

A = ordinary curve of gastric secretion.
B = curve from direct introduction of food into stomach.

demonstrated by comparing gastric digestion in which it is allowed to play a part, with gastric digestion in which the psychic element is shut out. This can be readily done in an œsophagotomized dog which possesses a fistula leading into the main stomach in addition to having a gastric cul-de-sac. If food is introduced into such a dog's main stomach without attracting its attention, it is found that digestion goes on at an entirely different rate than when the psychic element is allowed to play a part. Under these circumstances, bread and coagulated white of egg do not yield

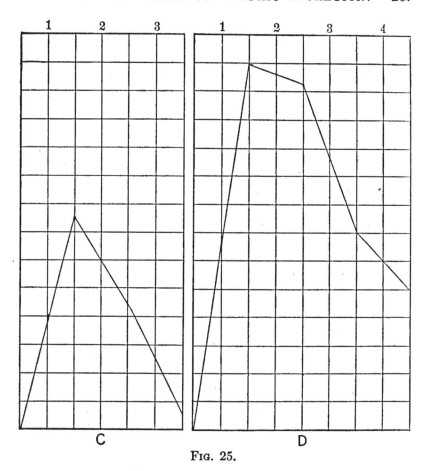

FIG. 25.

·*C* = curve from sham feeding.
D = summation of *B* and *C*.
(Copied from PAWLOW: Work of the Digestive Glands. Trans. by
THOMPSON, London, 1902, p. 82.)

a single drop of juice for an hour or more afterward and the
introduction of meat yields a gastric secretion only after a
considerable period of delay. It begins in the latter case in
15 to 45 minutes after feeding, instead of 6 to 10 minutes as
normally, and when it does flow is poorer, both in quantity
and quality, than under normal circumstances. In the accompanying figure (Fig. 25), curve *A* shows the ordinary course
of gastric secretion following a meal of 200 gms. of meat;

curve *B* that obtained when the same amount of food is introduced into the stomach directly, and curve *C* the result when sham feeding with the same is indulged in. *D* is a synthetic curve produced by adding *B* and *C* and approximates closely curve *A*. As can readily be seen, the introduction of food into the stomach directly does not lead to as rapid or as great a secretion of gastric juice as does the normal process of feeding. But if the quantities obtained by introducing the meat into the stomach be added to those obtained by sham feeding a curve almost identical with the normal results.

The importance of the psychic element in gastric digestion can be demonstrated still more strikingly by the following experiment of LOBASSOFF on two œsophagotomized dogs possessing ordinary gastric fistulæ. If a definite number of pieces of meat threaded on strings be introduced directly into the stomach of one dog while its attention is being distracted by patting and talking kindly to it, and an equal number of pieces are introduced into the stomach of another dog, but during the process a vigorous sham feeding is kept up, the following differences are noted. Two hours after 25 pieces weighing 100 gms. in all had been fed to each of the dogs, the dog without sham feeding had digested 6.5 percent, the one with sham feeding for eight minutes, 31.6 percent, as shown by subsequent weighing of the undigested food extracted from the stomach. In another experiment, in which the meat had remained for five hours in the stomachs, 58 percent had been digested without sham feeding, while 85 percent had been digested with sham feeding. These figures leave no room for doubt of the great importance to be attached to the eager desire for food, in other words to the appetite.

 6. **Other Excitants of Gastric Secretion.**—It must not be believed from the foregoing that the appetite is the only excitant of gastric secretion. This is shown not only by the observation that a secretion of gastric juice occurs when the psychic element is allowed to play no part at all, as

when the food is introduced into the stomach directly, but also by the fact that the effect of sham feeding does not last as long as an ordinary digestive period nor yield in the later hours of this period as active a juice as is obtained after a true meal. What, then, are the other excitants of the gastric flow besides the appetite which we have seen above plays so important a rôle in digestion? We think naturally of the mechanical stimulation and secondly of the chemical stimulation which might be brought about by the presence of the food in the stomach. Let us first see if mechanical stimulation is effective in bringing about a secretion of gastric juice from the stomach.

Almost without exception the older observers believed that mechanical stimulation of the gastric mucosa by a feather or a glass rod would yield at least some secretion of gastric juice. Within recent years, however, this question has been investigated by PAWLOW and his coworkers, who have pointed out the sources of error in the older experiments, and today we say that *mechanical stimulation is ineffective in bringing about a secretion of gastric juice.* This is shown by the following experiment.[1]

In an œsophagotomized dog possessing a gastric fistula the stomach is first thoroughly washed out with water. If the mucous membrane is now mechanically stimulated by moving a feather or a glass rod over it continuously for a half hour or more, or if a stream of fine sand is blown against it, or finally, if the stomach is distended to the size of a child's head by inserting within it a rubber ball, not a single drop of gastric juice is discharged. Only a little mucus which turns red litmus paper blue may be expelled. But let sham feeding with bread or meat be carried out upon the same dog, and an acid juice appears within five or ten minutes after the beginning of the feeding. The older observers never obtained

---

[1] PAWLOW: Work of the Digestive Glands. Translated by THOMPSON, London, 1902, p. 86.

through mechanical stimulation a gastric juice which even approximated the acidity obtained by sham feeding. That they obtained any acid secretion at all is due to errors in experiment, such as not washing out the stomach properly, not waiting until the stimulation to gastric secretion from the previous meal had entirely worn off, or causing a psychic secretion of the gastric juice by exciting the dog through the smell of food on the hands, the appearance of the attendant who usually fed the dog, etc.

Having settled now that the mechanical properties of the food are in themselves unable to call forth a secretion of gastric juice we turn to its chemical properties. What chemical constituents of the food bring about a secretion of juice from the stomach? In order to investigate this problem a dog with a miniature stomach, and possessing in addition a fistula passing into the main cavity, can best be used. The food undergoing study is introduced into the main cavity, and the amount and quality of the juice which flows from the isolated cul-de-sac examined. In order not to bring about a psychic secretion of the gastric juice, it is best to introduce the food while the dog is asleep, or while the dog's attention is distracted from what is going on if the animal is awake.

• *Water*, first of all, has an exciting effect upon the gastric glands. It has been found by CHIGIN that when 400 to 500 c.c. of water are introduced into the large stomach of a dog, a small but constant secretion of juice occurs from the lesser one. This fact had been previously found by HEIDENHAIN. Smaller quantities of water are not so effective, and if only 100 to 150 c.c. are injected into the large stomach, usually no secretion of juice at all occurs. The results are the same when before the introduction of the water the vagi nerves are divided below the diaphragm or in the neck.

Neither *solutions of sodium chloride, sodium bicarbonate,* or *hydrochloric acid* excite a flow of gastric juice. According to CHIGIN 0.05 to 1 percent sodium bicarbonate solutions, when introduced in the same amounts as proved effective when water alone was injected, bring about not even a slight secretion of gastric juice. This salt is, therefore, to be looked upon as *inhibiting* the gastric flow, for its presence prevents the usual exciting effects of the water alone. *Oils* also have a distinct inhibitory effect, according to LOBASSOFF.

Uncoagulated white of egg, whether diluted with water or not, never brings about a greater secretion of gastric juice than the same volume of pure water. This is an altogether unexpected fact, and points to the importance of the appetite juice in normal digestion. Neither does *starch*, boiled or unboiled and variously diluted, nor *grape-sugar*, nor *cane-sugar*, excite the stomach to secrete juice. Even *bread* and *boiled white of egg* remain for hours in the stomach and excite no gastric secretion if the psychic element is shut out.

Certain commercial peptones will excite the gastric mucous membrane directly, but by no means all of them. *Pure* peptones do not, however, have this effect. We are in consequence driven to the conclusion that the commercial peptones contain as yet unknown chemical substances which are the real excitants of the gastric flow. *Meat broth, meat juice,* and solutions of *meat extract* act as constant and active excitants of the gastric secretion. *Meat* also belongs in this group, but all these substances bring on a flow of gastric juice much later than when sham feeding is practised. Whereas the latter method shows a beginning of secretion after 5 minutes, the former is without effect for 15 to 45 minutes afterward. The majority of foodstuffs does not, therefore, affect the secretion of gastric juice. To the minority which is active in this direction, water and certain as yet unknown constituents of meat belong. The manner in which these are effective is now to be discussed.

7. Gastric Secretin.—When we study the quantitative variations in the secretion of gastric juice, we note that the curve of secretion shows two maximal points. The first of these is observed immediately after the ingestion of food, the second two or three hours later. The rapid secretion which. occurs at first is found even when the food never really enters the stomach, as in sham feeding. Since this secretion is associated with the passionate desire for food, or with the mental impressions produced by the sight, smell, taste, etc., of the food, it is termed the "psychic" element in gastric secretion. The channel over which the mental stimulation reaches the stomach and excites the latter to secretory activity is represented by the vagus nerves. When these are cut the initial rise in gastric secretion does not occur. The second great rise in the curve of gastric secretion following an ordinary meal nevertheless occurs. In what way is this second rise brought about?

The observations of EDKINS [1] give us an answer to this question. Familiar with the experiments of BAYLISS and STARLING on pancreatic secretin,[2] EDKINS cast about to find a gastric secretin. His experiments show that *under the influence of certain digestion products, the mucosa of the pyloric end of the stomach produces, during the period of digestion, a substance—gastric secretin—which is absorbed into the blood and which excites the gastric glands to increased secretion.* To prove this EDKINS prepared extracts of the mucous membrane of the pyloric end of the stomach by rubbing this up in a mortar with 5 percent dextrin solutions, or solutions of various sugars and peptones. If repeated small doses of this extract are injected into the circulation of a dog which has had its stomach filled with a physiological salt solution, it is found at the end of an hour that the salt solution contains both hydrochloric acid and acid proteinase. A secretion of gastric juice is therefore excited by these extracts. It can be shown in control experiments in which dextrin, peptone, etc., are injected in pure solution into the circulation that no hydrochloric acid or acid-proteinase can be found in the physiological salt solution recovered from the stomach.

[1] EDKINS: Journal of Physiology, 1906, XXXIV, p. 133; STARLING: Lancet, 1905, CLXIX, p. 501.

[2] See p. 227.

We may conclude, therefore, that the increased secretion of gastric juice, brought about through the injection of extracts of the mucosa of the pyloric end of the stomach, is due to the injection of some substance which is produced in this region of the stomach under the influence of certain of the products of gastric digestion. This substance is called gastric secretin. The mucous membrane of the fundus of the stomach does not contain this substance. As to the nature of gastric secretin but little can be said. It does not seem to belong to the class of ferments, for it is not destroyed by boiling. In this regard it is similar to pancreatic secretin.

If now, in the light of the experimental facts which have been cited above, we try to explain the progress of normal gastric digestion the following may be said: The initial secretory period, which is noted after an ordinary meal eaten in the ordinary way and with desire, is explained by the psychic effect of eating. This psychic effect lasts for three or four hours. The digestion periods following this are independent of the central nervous system and are governed by chemical agents. In the case of meat we find a reason for the continued secretion of gastric juice after the initial psychic period in the chemical constitution of the food itself. It contains substances which, acting on the mucosa of the pylorus, cause the elaboration of gastric secretin. The same holds true for predigested foods—that is, foods containing digestion products which act in a similar way upon the pyloric mucosa. In the case of the remaining substances, such as bread, white of egg, etc., we can say that under normal circumstances the psychic juice starts their digestion, and in this process chemical substances are formed which bring about an elaboration of secretin, and this keeps up the gastric flow after the psychic element has come to rest. This idea is supported by the fact that the products of digestion formed in the stomach of one dog act as excitants of gastric secretion when introduced into the stomach of another. The clinical application which may be made of these experimental facts is too evident to need much comment. The experiments indicate very clearly that lack of appetite means lack of gastric juice, which, in turn, is synonymous with faulty gastric digestion. We see also the usefulness of predigested foods in certain pathological conditions of the stomach and the rational basis for the use of soups, meat extracts, etc., which, though they have no food value in themselves, are useful remedial agents, since they cause a secretion of gastric juice which may be utilized in digesting otherwise indigestible constituents of a mixed meal.

CHAPTER XII.

THE REGULATION OF THE PANCREATIC SECRETION.

1. Pancreatic Fistulæ.—For the study of the quantitative and qualitative variations in the secretion of the pancreas under different physiological conditions various experimental procedures have been adopted from time to time. The earlier observers contented themselves with isolating and dissecting out the pancreatic duct, inserting a cannula into it, and collecting the juice which flowed from it. It was soon found, however, that the anæsthetic, surgical shock, etc., so affected the activity of the gland as to stop its secretion altogether, or at the best allow the flow of only a small amount, and that not very active pancreatic juice.

In endeavoring to overcome the objections against such a "temporary" fistula of the pancreas, Claude Bernard and Ludwig attempted to produce a "permanent" one which should be free from the immediate effects of an operation. The former observer tied a glass cannula into the secretory duct of the pancreas and brought it out through the abdominal wall; the latter used a lead wire to keep the duct patent. The improved technique did, in fact, yield better results than the older methods, but in the course of five to ten days the cannula or wire sloughed out, and further experimentation was interfered with through infection of the operation wound and pancreas.

In 1879 Pawlow[1] and a year later Heidenhain[2] described

---

[1] Pawlow: The Work of the Digestive Glands. Translated by Thompson, London, 1902, p. 5.

[2] Heidenhain: Hermann's Handbuch der Physiologie, Bd. V, p. 177.

a surgical procedure for the production of a "permanent"
pancreatic fistula which has stood the test of time, and upon
which our modern knowledge of the activity of the pancreas
is largely based. PAWLOW's method is as follows: An oval
piece containing the orifice of the pancreatic duct is cut out of
the duodenum, and the opening in the intestine is closed by a
row of sutures. The isolated piece of intestine is then carried
up into the wound in the abdominal wall and sutured there
with the mucous membrane directed outward. At the end
of two weeks the animals may be used for study. When the
wound has healed, it shows a roundish elevation of the mucous
membrane, in the centre of which the cleft-like orifice of the
pancreatic duct is clearly visible. By supporting the animal
in a suitable frame the pancreatic juice may now be collected
directly into a graduated cylinder as it drops from the duct,
or if the juice tends to spread over the abdominal wall a funnel
may be strapped over the opening of the duct. If only care
be taken to keep the fistulous opening clean, to avoid macera-
tion of the skin by allowing the animal to lie in sand or saw-
dust which absorbs the pancreatic juice, and to supply the
animal with proper food, the dog operated upon in the way
indicated may be kept for months and even years in a healthy
condition.

More recently the operation of PAWLOW and HEIDENHAIN
has been much improved by FODERÀ[1] who succeeded in causing a T-shaped metallic cannula to heal into the pancreatic duct.
In this way the pancreatic juice may be collected either externally or, by closing the outer opening, be diverted into
the lumen of the intestine. Thus the deleterious effects of
the constant loss of pancreatic juice, necessary in PAWLOW
and HEIDENHAIN's operation, is prevented during those periods
when the animal is not being used for experimental purposes.

2. **The Effect of Diet on Pancreatic Secretion.**—In a dog
in which a pancreatic fistula has been created by either the

---

[1] FODERÀ: Moleschott's Untersuchungen zur Naturlehre d. Menschen
und d. Tiere, 1896, XVI.

method of PAWLOW or FODERÀ, as described above, the quantitative and qualitative variations in the pancreatic juice under various conditions of diet, etc., can be followed with great exactness. As was found to be true in the case of the stomach, the secretion of pancreatic juice is intimately connected with the taking of food. The pancreatic secretion, which, during fasting, amounts to only two or three cubic centimeters in twenty-four hours, is increased to many times that amount as soon as food is taken. The secretion from the pancreas is poured out gradually, though different amounts enter the intestine in each unit of time. The following experiment of WALTHER [1] illustrates this point.

RATE OF PANCREATIC SECRETION AFTER FEEDING 600 C.C. MILK.

Hour after feeding.	Quantity of juice in c.c.	
	Exp. *a.*	Exp. *b.*
1	8.75	8.25
2	7.5	6.0
3	22.5	23.0
4	9.0	6.25
5	2.0	1.5

These values are expressed graphically in Fig. 26.

Not only does the *quantity* of pancreatic juice secreted from hour to hour vary but also the *quality*. In the following table are indicated the hourly variations in the digestive power of pancreatic juice after a meal of 600 c.c. milk. The digestive power of the fat-splitting ferment is expressed in terms of cubic centimeters of a standard barium hydrate solution, required to neutralize the acid formed in a given length of time, when the specimen of pancreatic juice is allowed to act on a fat. The activity of the proteolytic ferment is determined by METT's method. That of the amylolytic ferment is determined by an analogous method, in which colored starch paste is digested out of capillary tubes. In the case of the proteolytic and amylolytic ferments, the

---

[1] WALTHER: PAWLOW's Work of the Digestive Glands. Translated by THOMPSON, London, 1902, p. 22.

activity is expressed in number of centimeters of egg-albumin
or starch digested out of the tubes in the unit of time. As
has already been pointed out, this method is not free from

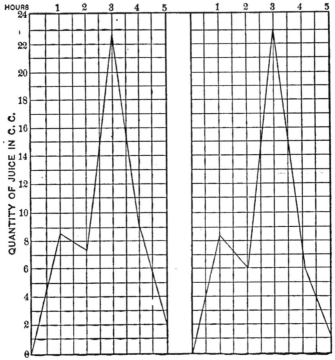

FIG. 26. Curve of secretion of pancreatic juice after feeding 600 c.c. .
milk. Two experiments.

(Copied from PAWLOW: Work of the Digestive Glands. Trans. by
THOMPSON, London, 1902, p. 24.)

objection. The values obtained in different experiments
agree so well with each other, however, that the results must
be looked upon as correct, at least in the main.

HOURLY VARIATION OF DIGESTIVE POWER OF PANCREATIC JUICE
AFTER A MEAL OF 600 C.C. MILK.

Hour.	Lipolytic ferment.		Amylolytic ferment.		Proteolytic ferment.	
	Exp. *a.*	Exp. *b.*	Exp. *a.*	Exp. *b.*	Exp. *a.*	Exp. *b.*
1	14.0	14.0	5.1	5.0	5.8	5.5
2	20.0	13.0	5.0	4.7	5.9	5.5
3·	7.0	5.2	2.4	2.4	4.3	4.1
4	6.0	7.0	3.3	3.4	4.5	4.4

These values are expressed in curves in Fig. 27.

The cause of these variations in the digestive power of the pancreatic juice is not yet explained. A strong digestive

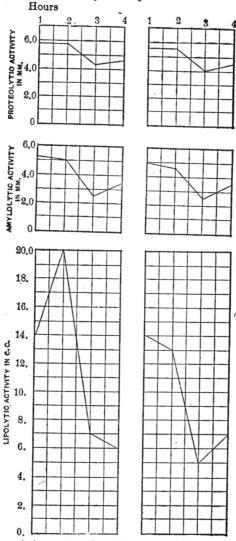

Fig. 27.—Fermentative activity of hourly portions of pancreatic juice after a meal of 600 c.c. milk.
(Copied from Pawlow: Work of the Digestive Glands. Trans. by Thompson London, 1902, p. 30.)

power may be found when much juice is being secreted by the gland as well as when the flow is scanty. The variations are probably associated with the progressive changes which the food undergoes during digestion, but just how they are connected is not yet known.

The variation, both in quality and quantity, which the pancreatic juice suffers when *different diets* are fed is indicated in the following table. By *strength of juice* is meant the square of the number of millimeters of egg-albumin or starch dissolved out of the capillary tubes, or the square of the number of cubic centimeters of standard barium-hydrate solution used to neutralize the acid formed from the digested fat. By the *total quantity of ferment units* is meant the product of the strength of the juice multiplied by the quantity of the juice in cubic centimeters. The amounts of food chosen represent equivalents of nitrogen (WALTHER).

Diet.	Quantity of juice in c.c.	Proteolytic ferment.		Amylolytic ferment.		Lipolytic ferment.	
		Strength of juice.	Total quantity of ferment units.	Strength of juice.	Total quantity of ferment units.	Strength of juice.	Total quantity of ferment units.
Milk, 600 c.c. ..	48	22.6	1085	9	432	90.3	4334
Bread, 250 gms.	151	13.1	1978	10.6	1601	5.3	800
Meat, 100 gms..	144	10.6	1502	4.5	648	25.0	3600

In discussing the gastric secretion and its variations under the influence of different diets, it was pointed out that not only the quantity but also the digestive power of the juice differs with different kinds of food. The above table shows that for the pancreatic juice it is still more strikingly true that each kind of food has its own particular kind of juice. Each sort of food determines the secretion of a definite amount of pancreatic juice. But still more remarkable is the variation in the amount of the different ferments poured out upon the foods. The juice poured out upon bread is exceedingly poor in lipolytic ferment, while that poured out upon milk is very rich. The juice poured out upon meat occupies an intermediate position in this regard. The largest amylolytic activity is found in the juice poured out on bread, less in the juice poured out on milk, and still less in that poured out upon meat. It is self-evident that so far as the fat- and starch-splitting ferments are concerned the properties of the pancreatic juice correspond to the requirements of the food. A diet rich in starch receives a juice rich in amylolytic ferment, one rich in fat a juice containing much lipolytic ferment. This is shown, as indicated in the above table, not only by the strength of juice poured out upon these different foods but also by the absolute quantities secreted by the gland.

The behavior of the proteolytic ferment seems at first to differ from that of the amylolytic and lipolytic. In discussing gastric secretion it was shown that the weakest juice was poured out on milk. In the case of the pancreas, however, the strongest proteolytic juice is poured out upon this food, while a weaker one is poured out upon bread and a still weaker one upon meat. When, however, the quantity of juice poured out upon these different foods is taken into consideration we arrive at conclusions similar to those found to hold for gastric secretion. As indicated in the above table, when equivalent quantities of protein are fed in the form of bread, meat, and milk, the *total quantity of ferment units* poured out upon these foods stands as 1978:1502:1085. In other words, vegetable protein demands from the pancreas, as well as from the stomach, the largest number of ferment units, while milk demands the least. The difference between the secretion of the stomach and the pancreas is, therefore, limited to the fact that the former pours out its ferment in a very concentrated form upon bread, while the latter pours it out in a more dilute solution.[1]

---

[1] See PAWLOW: Work of the Digestive Glands. Translated by THOMPSON, London, 1902, p. 39.

3. **The** Relation of the Nervous System to Pancreatic
Secretion.—The secretion of juice by the pancreas is influenced by the vagus nerve in much the same way as the
secretion from the stomach. In this chapter of physiology
also, PAWLOW and his coworkers have done most within recent
years to advance the state of our knowledge. The following
experiment shows unequivocally that the vagus nerve contains fibres which influence the secretion of the pancreatic
juice. A dog in which a permanent pancreatic fistula has
been made after the fashion described above is employed. The
vagus nerve is divided in the neck on one side, the peripheral
end is laid bare, and after having a ligature passed around
it, is preserved under the skin. On the fourth day after
division the severed nerve is carefully pulled out of the
wound without hurting the dog in any way. No juice flows
from the pancreatic fistula. But one or two minutes after
the peripheral end of the vagus is stimulated by an induction current a drop of juice appears at the orifice of the
pancreatic duct, which is soon followed by another and
another in rapid succession. If the induction current is
interrupted the pancreatic juice continues to flow for four
or five minutes and then ceases. If now the stimulation be
renewed juice appears a second time, and so on.

The reason why the older observers never obtained a flow
of pancreatic juice when they stimulated the vagus is due to
the fact that in their experiments performed at one sitting
peripheral stimuli initiated by anæsthetic, pain due to operation, etc., acted upon the pancreas and produced a reflex
inhibition. That such sensory stimuli exert an inhibitory effect upon the pancreatic gland has been shown by BERNSTEIN,
PAWLOW, and others. Circulatory disturbances also interfere with pancreatic activity. In the experiment described
above such peripheral stimuli are carefully avoided, first
by utilizing an animal which has recovered from the effects
of anæsthetic and operation, and secondly by stimulating
the vagus at a time when its cardiac fibres have degenerated

to such an extent that stimulation no longer affects the heart. At this time the fibres influencing the pancreatic secretion are, however, still active.

But the influence of the vagus nerve can also be demonstrated in an experiment performed at a single sitting. Special precautions must be taken, however, to prevent the inhibition of the gland either through nervous reflexes or through the circulation. To do this a dog has his spinal cord divided just below the medulla in order to shut out all peripheral stimuli which might inhibit the activity of the pancreas. After this has been done and artificial respiration started, the thorax is opened and the vagus divided below the heart. If now a cannula is inserted into the pancreatic duct a secretion of pancreatic juice is noticed whenever the peripheral end of the cut vagus is stimulated. By thus exciting the vagus in the thorax its influence on the heart is avoided.

The experiments of POPIELSKI [1] indicate that the vagus nerve contains fibres which not only augment the secretion of the pancreas but also such as inhibit it. If during the acute experiment described above a solution of hydrochloric acid is poured into the duodenum a vigorous secretion of pancreatic juice is called forth which continues for a long time. If now the vagus nerve be stimulated the secretion is markedly diminished in amount, often stopped entirely.

The pancreas is influenced also by the sympathetic nerve, as shown by the experiments of KUDREWETZKY.[2] This nerve apparently carries two kinds of fibres to the gland—first, vaso-constrictor, and secondly, such as influence the secretory activity of this organ. If the sympathetic is stimulated in the ordinary way by electric induction shocks a slight increase in secretion is observed, which lasts, however, only a few seconds; then it ceases entirely even if the electrical stimulation is continued. This is due to the fact that the

---

[1] POPIELSKI: Centralbl. f. Physiol., 1896, X, p. 405.

[2] KUDREWETZKY. Archiv f. (Anat. u.) Physiol., 1894, p. 83.

vaso-constrictor effects completely mask the secretory. The two can be separated by mechanical stimulation of the sympathetic. When this is done stimulation always leads to an increased secretion from the gland, for, as is well known, vaso-constrictor nerve fibres are not readily excited by mechanical means. The two can also be separated by stimulating the sympathetic several days after it has been divided. The vaso-constrictor fibres are the first to degenerate, so that only those influencing the secretory activity of the gland are left to exhibit their characteristic effects.

4. The Normal Excitants of the Pancreas.—It has been shown above that during a period of digestive inactivity the pancreas secretes no juice; that after feeding, a large amount of pancreatic juice is secreted. What determines this secretion? The relation of the various inorganic constituents of the food to the pancreatic flow has been studied by BECKER [1] and DOLINSKI.[2]

Acids of all kinds act as powerful excitants of the pancreatic secretion. If 250 c.c. of a 0.5 percent *hydrochloric acid solution,* for example, are introduced by means of a catheter into the stomach of a dog possessing a permanent pancreatic fistula, the inactive pancreas begins to secrete within two or three minutes. Within ten minutes the pancreatic flow reaches its maximum and within the first hour some 80 c.c. of juice may be collected. The rate of secretion steadily falls, so that towards the end of the second hour only about 15 c.c. can be collected before the outflow ceases entirely. If, instead of the hydrochloric acid solution, an equal amount of water is injected into the stomach little or no pancreatic juice is obtained, which shows that it is the acid which is active. That at the end of such an experimental period the pancreas is not exhausted can be readily proved by injecting

---

[1] BECKER: Archives des Sciences Biologiques, II, p. 433. PAWLOW'S Work of the Digestive Glands. Translated by THOMPSON, London, 1902, p. 113.

[2] DOLINSKI: Arch. des Sci. Biolog., III, p. 399.

the hydrochloric acid a second time, when figures practically the same as those given above can again be obtained. *Phosphoric, citric, lactic,* and *acetic acid* behave in a way similar to hydrochloric. *Carbonic acid* also belongs in this group.

Sodium chloride in solution seems to be without effect upon the pancreas, and *alkaline solutions,* such as calcium hydroxide, sodium bicarbonate, and alkaline mineral waters, have a distinctly *inhibitory* effect upon its secretion.

After what has been said it is not strange that gastric juice acts as a chemical excitant of the pancreas, and to the same degree as a hydrochloric acid solution of the same concentration. When the gastric juice of the acid food passing out of the stomach is neutralized the effect of the latter in calling forth a pancreatic secretion disappears. In what way the acid is effective in acting upon the pancreas will be discussed later when we speak of *secretin.*

In spite of the fact that *starch* is one of the constituents of the food upon which the pancreatic juice acts, this carbohydrate, whether boiled or unboiled, and at any concentration, affects the quantity of juice poured out by the pancreas no more than an equal amount of water. A qualitative change in its composition is, however, rendered probable by the experiments of WALTHER, who found that a dog when fed on bread secretes a pancreatic juice richer in amylolytic ferment than when fed with meat.

Fat seems to be an excitant of the pancreas. DOLINSKI and DAMASKIN [1] found that after introducing oil into the stomachs of dogs a flow of pancreatic juice always ensued.

Do we have a *psychic secretion of pancreatic juice* similar to the psychic secretion of gastric juice? This question cannot yet be looked upon as definitely settled. KUWSCHINSKI [2] showed in 1888 that tempting a hungry dog with food led

---

[1] PAWLOW: Work of the Digestive Glands. Translated by THOMPSON, London, 1902, p. 121.

[2] KUWSCHINSKI: PAWLOW's Work of the Digestive Glands. Translated by THOMPSON, London, 1902, p. 121.

to a free secretion of pancreatic juice.   This does not prove, however, that the pancreas can be directly excited by this means, for to tempt a hungry dog is to excite a gastric secretion, and the acid flow produced in this way might readily be the real excitant of the pancreatic flow.   Two facts, however, point strongly in favor of the existence of a psychic secretion of pancreatic juice.   First, tempting a dog with food or sham feeding will cause a quiescent pancreas to become active even when a gastric fistula allows the escape of the gastric juice from the stomach as soon as formed; secondly, the latent period marking the beginning of secretion from the stomach and the pancreas is different in the two cases.   An acid flow begins from the stomach five to ten minutes after the beginning of sham feeding.   The ,pancreas begins to secrete after two to three minutes.   If the acid and not a true psychic element were responsible for the secretion of pancreatic juice the pancreas should not become active until after the gastric flow has commenced.

Water also acts as an excitant of the pancreas.   When 150 c.c. of water are introduced unnoticed into the stomach of a dog the pancreas begins to secrete, or augments its flow two or three minutes after the water has entered the stomach. If the experiment is properly performed this amount of water does not excite a flow of gastric juice, so that a secretion of pancreatic juice secondary to a secretion of gastric juice is out of the question here.

Meat extracts, which were found above to be such good excitants of the gastric glands, are no more effective in the case of the pancreas than an equal amount of water.

What has been said in this section regarding the influence of acids, food, etc., on the qualitative and quantitative composition of the pancreatic juice in the dog seems to hold without modification for the human being.   As evidence in this direction we may cite the experiments of GLAESSNER [1] on

---

[1] GLAESSNER: Zeitschrift für physiologische Chemie, 1904, XL, p. 465

his patient with a fistula of the pancreatic duct. GLAESSNER found that the *amount* of juice secreted in the unit of time was greatly increased soon after his patient took food. Whereas under ordinary circumstances 10 to 15 c.c. flowed from the fistula per hour, 38 c.c. were obtained in the first hour after a meal, 34 c.c. in the second, and 46 c.c. in the fourth. From this hour on the flow steadily decreased until at the end of the eighth it had fallen to its original level. It is evident, therefore, that the curve representing the rate of secretion in the human being is similar to that which has been described for dogs.

The amount of lipolytic, amylolytic, and proteolytic *ferment* in the unit volume of pancreatic juice GLAESSNER also found to vary with the taking of food. Between meals this was least, while the slow rise which began during the first or second hour after eating was found to attain its maximum about the fourth, when it fell once more in the following four hours to its original level. The *alkalinity* of the juice was also found to increase after the taking of food. While with phenolphthalein as indicator 1 c.c. of decinormal acid was required to neutralize 10 c.c. of pancreatic juice between meals, it required 5 c.c. of the acid to neutralize the same amount of juice obtained during the fourth hour after a meal.

The presence of an acid in the duodenum seems to increase the pancreatic flow in a human being just as in a dog. GLAESSNER found that the introduction of several hundred cubic centimeters of a weak hydrochloric acid into the stomach of his patient was followed almost immediately by an increased pancreatic flow which attained its maximum by the end of the first hour, and fell to the original level once more by the end of the second, as in the case of the dog.

5. **Pancreatic Secretin.**—In discussing the effect of acids upon the pancreatic flow, no mention was made of the mechanism by which this most powerful excitant of the gland accomplishes its results. PAWLOW believed that the gland was excited through a nervous reflex, initiated by the effect of

the acid upon the mucous membrane of the duodenum. That this explanation is incorrect is indicated by the experiments of POPIELSKI, WERTHEIMER, and LEPAGE, who found that the introduction of acid solutions into the stomach or duodenum still excited the gland to secretion, when all the nerves supplying the pancreas and duodenum had been cut. These authors, nevertheless, clung to the idea of a nervous excitation of the gland, at least in part. What had been rendered probable through the experiments of POPIELSKI, WERTHEIMER, and LEPAGE was made still more evident by the researches of BAYLISS and STARLING, who found that it is possible to isolate the pancreas from the duodenum, to divide all the nerves going to a loop of the small intestine, and nevertheless get a copious pancreatic secretion soon after an acid is injected into this loop.

How, then, does the acid act? The experiments of BAYLISS and STARLING [1] have shown that the connection between intestine and pancreas is really only a chemical one. The introduction of an acid into the duodenum or upper part of the small intestine brings about the production of a chemical substance in the mucous membrane which is called *secretin*. Since there are other secretins, it is well to call the one under consideration here *pancreatic secretin*. As the pancreatic secretin is formed it is rapidly absorbed into the blood, with which it travels to the pancreas, and excites this organ to secretory activity.

Pancreatic secretin may be obtained from the upper part of the intestine of any of the vertebrates. It is best prepared by scraping off the mucous membrane of the upper two feet of the small intestine, treating this in a mortar with sand and 0.4 percent hydrochloric acid, and then boiling the mixture. After neutralization with caustic soda the whole is filtered. The clear liquid which passes through the filter contains the secretin. In order to obtain the secretin in a

[1] BAYLISS and STARLING: Proceedings of the Royal Society, 1904, LXXIV, p. 310; STARLING: Lancet, 1905, CLXIX, pp. 339, 423, 501.

still purer state, the filtrate may be treated with absolute alcohol and ether, which precipitates certain impurities but allows the secretin to remain in solution. Secretin is therefore a substance which is unaltered by boiling and is soluble in alcohol. It can moreover be shown that it is diffusible. It is not necessary that the mucous membrane from which the pancreatic secretin is to be obtained should be living. It can be obtained just as well from an intestine which has been killed by boiling. Pancreatic secretin has, therefore, none of the ordinary characteristics of a ferment, and recent experiments seem to indicate that it belongs to the class of the peptones. It is possible, therefore, that secretin represents nothing but a product of proteolytic activity.

A few cubic centimeters of the secretin-containing filtrate obtained as described above when injected into the veins of an animal call forth a plentiful secretion of clear pancreatic juice, which may be collected by means of a cannula placed in the pancreatic duct. The pancreatic secretin obtained from one animal is not specific for that class alone, but will when injected into the veins of any other vertebrate call forth a copious pancreatic flow. It is well to add that the pancreatic juice obtained after the injection of secretin is to all appearances identical with that obtained after an ordinary meal.

Bayliss and Starling's findings alter somewhat our conception of the events which follow each other in the intestinal canal. A consideration of the facts which have been outlined above makes us imagine the changes which occur here to be somewhat as follows: The acid gastric contents enter the duodenum and bring about the formation of pancreatic secretin. This is absorbed into the circulation and, passing with the blood-current through the pancreas, causes a secretion of pancreatic juice. Since the pancreatic juice is able to neutralize free acids, it will decrease little by little the acidity of the gastric contents which have come into the duodenum. Little by little also will the rate of production of secretin be decreased. But not until all the acid in the duodenum has been neutralized will the production of pancreatic secretin, and, in consequence, the pancreatic flow, cease entirely.

This chemical interaction between the stomach, duodenum, and pancreas is further aided by the rhythmic opening and closing of the pyloric sphincter, which has already been described.[1] The presence of free hydrochloric acid in the stomach causes the pyloric sphincter to relax and some of the gastric contents to pass into the duodenum. As soon as the acid reaches this part of the intestinal canal, however, the sphincter is closed reflexly, and remains closed until the duodenal contents have become neutralized by being mixed with pancreatic juice. As soon as this neutralization has been accomplished, the pylorus relaxes a second time and a fresh portion of the gastric contents enters the duodenum. This in its turn calls forth a production of pancreatic secretin, a secretion of pancreatic juice, and the cycle is repeated.

6. Significance of the Physiology of the Gastric and Pancreatic Secretions.—The advances made in our knowledge of gastric and pancreatic activity have done much to give us an experimental basis for what has hitherto been empirical in medical practice and have at the same time pointed out the direction which medicine must take in combating certain affections of the alimentary tract. The experimental results detailed above show most clearly the physiological importance of the appetite. Appetite is synonymous with a copious secretion of gastric juice, and the necessity of this secretion for a proper and rapid digestion of proteins and the evil effect of its absence are manifest nowhere more clearly than in those disorders which are associated with a deficient secretion from the stomach. We recognize, in consequence, the importance of catering to the appetite in health, in order to maintain gastric digestion at its best, and the urgency of restoring this physiological sense in those diseases in which

---

[1] See p. 16.

it is absent. As the pancreatic secretion is probably in_
fluenced by appetite in the same way as the stomach, we see
an additional reason for recognizing its medical importance.
The influence of the appetite upon the secretion of the gastric
juice contributes also to our understanding of the beneficent
effects which follow the feeding of a patient with a "weak"
stomach at frequent intervals and only small amounts at a
time. Under these circumstances the appetite juice, rich in
ferments, recurs several times and digestion is furthered in
consequence.

We are also in a position to understand now the good effects
of condiments and bitters. Pharmacologists have looked in
vain for a direct secretory effect of these substances upon
the stomach and pancreas. Every-day observation, however,
seems to support indisputably the fact that these substances
increase the appetite. If this be true, we have arrived at an
explanation of the good effects following their moderate use,
for to increase appetite is to increase gastric and pancreatic
secretion.[1] We can recognize also the relation which appetite
and gastric secretion bear to each other. Quite contrary
to the generally accepted idea that the presence of gastric
juice in the stomach is the cause of appetite, we must say that
gastric secretion is its consequence.

It is necessary to revise our ideas of the usefulness of meat
soups, meat extracts, meat juices, etc. Few medical men to-
day are not acquainted with the valuelessness of the first two
from a nutritive standpoint. But since they excite a secretion of gastric juice, even in the absence of appetite, they are
eminently useful, for they can in this way be employed to
bring about the digestion of food which is introduced into
the stomach subsequently. Because of this fact soups ful-
fil a good function when employed at an ordinary meal, for
they help to maintain a secretion of digestive juice which has
been inaugurated by the appetite.

---

[1] See PAWLOW: Work of the Digestive Glands. Translated by THOMP-
SON, London, 1902, p. 138 et seq.

Milk and water are also independent excitants of the gastric secretion. We find in this another reason for the use of water with meals. Milk has from the earliest times been considered an easily digested food for children and invalids. Not only does this fluid cause a secretion of gastric juice, but, as has been shown above, a weaker gastric juice and a smaller quantity of pancreatic juice are poured out on milk than on an equivalent amount of nitrogen contained in any other food. The work performed by these digestive glands is therefore less, and the saving of energy, in consequence, is greater than when meat or bread, for example, is fed. The amount of work which the organism does in order to assimilate a certain amount of nutriment must always be subtracted from the energy value of the food itself. This is not ordinarily considered in determining the value of various foods. We have seen above, however, that equivalent weights of nitrogen in the proteins of meat, bread, and milk cost the organism different amounts of glandular energy.

Fat inhibits the secretion of gastric juice. We find in this the explanation of the fact that its presence in a mixed meal impedes the rate of digestion, and hence is contraindicated in cases of feeble digestion, while it works excellently in cases of hyperacidity and hypersecretion. The alkalies and alkaline salts belong in this group of substances which inhibit the secretions of the gastric mucosa and the pancreas.

We have learned from experiments detailed above the uselessness of mechanical stimulation of the gastric mucosa in bringing about a secretion of gastric juice. Neither mechanical stimulation through coarse food or by means of special apparatus, such as an inflatable balloon, can therefore be looked upon as rational treatment for those gastric disorders which are characterized by deficient secretion of juice. Care must be taken not to confound this statement with the relation between mechanical stimulation and the *motor* functions of the stomach.

A defective secretion of gastric juice, be its cause what it may, is of medical importance from other points of view than that it interferes with the proper digestion of proteins. A lack of hydrochloric acid in the stomach allows the develop- ment here of a large number and variety of bacteria which in its presence is impossible. The development of these bacteria is associated with fermentative changes in the stomach contents, which assume clinical importance in proportion to their character and amount. Lack of free hydrochloric acid in the stomach delays the opening of the pyloric sphincter, which leads to retention of the food in the stomach for longer periods of time than normally. In this way also bacterial fermentation is allowed to become more effective.

When ultimately the stomach contents pass into the small intestine, they flood this portion of the alimentary tract with bacteria and, as already pointed out, the bactericidal activity of the small intestine is by no means limitless. The fermentative changes in the food initiated in the stomach may in consequence continue in the intestine, with effects upon the organism as a whole (due in part to absorption of the bacterial products formed, in part to a lack of properly digested foods) which are sometimes unimportant, sometimes serious.

A deficient amount of gastric juice leads to yet other disturbances in the organism. We are slowly beginning to recognize the intimate connection which exists between different organs in the body and how the proper behavior of one is dependent upon that of another. This relation between organs exists also between different portions of the alimentary tract. It might be thought at first that from the standpoint of digestion alone a deficient amount of gastric juice is without consequence, for the pancreas contains a proteolytic ferment which is able to split proteins even more readily than the acid-proteinase of the stomach. Matters are not so simple, however. It was seen above that the presence of acids in the duodenum brings about a secretion of pancreatic juice. If now the gastric secretion is de-

ficient, it means a lack of acid in the duodenum, and this
in turn means a deficient secretion of pancreatic juice.

**7. On the " Adaptation " of the Digestive Glands to the
Character of the Food.**—The opinion that the secretions
of the various digestive glands "adapt" themselves to the
character of the food consumed by the individual, that, in
other words, each diet calls forth a particular kind of digestive secretion, has been expressed from time to time by various
authors.   With all of them, however, this opinion has rested
more upon philosophical speculations than upon experimental
facts.

The first to attempt to give experimental foundation to such
an idea was PAWLOW, and with him CHIGIN and WALTHER.
But the experiments which these investigators have detailed
in support of their belief, and which have been touched upon
in the preceding pages, have been so energetically attacked
by others that, at the best, they can be looked upon as by
no means convincing.

With all the more pleasure, therefore, can we proceed to a
discussion of the experiments of WEINLAND,[1] which indicate
without question that one digestive gland at least, the pancreas, can and does "adapt" itself to a particular diet.
WEINLAND's experiments show that the pancreas and presumably, therefore, the pancreatic juice of the adult dog,
which normally contain little or no lactase, come to contain this enzyme in large amounts if the dog is fed milk-
sugar, or a diet containing milk-sugar, such as milk.   This
means that the pancreas and pancreatic juice, which in the
adult dog normally contain little or no ferment which can act
upon lactose and split this into galactose and dextrose, develop this sugar-splitting ferment in response to certain diets.

What has been said can be best illustrated by introducing
one of WEINLAND's experiments.   An adult dog was kept on
an abundant diet, but free from milk-sugar, for five days;

---

[1] WEINLAND: Zeitschr. f. Biol., 1899, XXXVIII, p. 16; ibid., 1899,
XXXVIII, p. 607; ibid., 1900, XL, p. 386.

he was killed and the pancreas removed immediately, an extract made of it and its power of acting upon milk-sugar (lactase content) determined. The extract was found to have practically *no effect* upon milk-sugar, indicating, therefore, that it contained little or no ferment capable of acting on lactose. A second, also *adult* dog received for 15 days a diet similar to that given the first dog, only 30 gms. of milk-sugar were added to the daily food ration. When at the end of this time the dog was killed, the pancreas removed and treated as that of the first dog, it was found that the *pancreatic extract split milk-sugar most energetically* into galactose and dextrose, 43 percent of the added lactose being split within twenty-four hours. It is self-evident, therefore, that under the influence of milk-sugar feeding the pancreas can develop a ferment (lactase) which has the power of splitting milk-sugar and which is entirely absent (or present in only very small amounts) in the pancreas of an adult dog not so fed.

The percent of sugar split as given above is by no means the highest that WEINLAND ever attained. In three other experiments in which the dogs were fed ordinary milk instead of pure milk-sugar, the pancreatic extracts were able to split respectively 54 percent, 73 percent, and 75 percent of the lactose added to them. When it is pointed out that the pancreas obtained from dogs of a corresponding age, but not fed on milk-sugar, possess practically no milk-sugar-splitting activity it is self-apparent how striking is this "adaptation" of the pancreas to the character of the diet.

It is of physiological interest that the quantitative variation in the amount of lactase found not only in the pancreas but also in the small intestine is intimately connected with the age of the individual, or, as we can say now, with those periods of life in which milk-sugar furnishes a part of the food consumed by the animal. Apparently all sucklings (including the new-born child) have lactase present in the pancreatic juice and the intestinal juice. This has been proven true for all mammals examined. In adult life, however, the lactase of the pancreas disappears very largely, at times entirely, and the lactase of the intestinal mucosa and its secretion decreases much in amount (and in some animals disappears entirely) unless milk-sugar continues to serve as an article of food for the adult animal.

We have yet to discuss the mechanism of this adaptation. A direct effect of the milk-sugar upon the pancreas from the lumen of the intestine is practically impossible, for the pancreas is connected with the gut by only a slender excretory duct, through which, moreover, during the periods of digestion an uninterrupted stream of juice pours. It might be thought that when an excessive amount of milk-sugar is fed, some passes over into the circulation, for we can discover milk-sugar in the urine; and this absorbed sugar might be thought to act directly upon the pancreas. That this idea is not correct follows from the fact that the intravenous or subcutaneous injection of milk-sugar is not followed by the appearance in the pancreas of an increased amount of lactase. Neither does a subcutaneous or intravenous injection of one of the *products* of milk-sugar digestion, such as galactose, accomplish this result. In order that this may occur *the milk-sugar must come in contact with the mucosa of the intestine.*

We are indebted to VERNON [1] for a further analysis of the problem. This author, who entirely confirms the experimental findings of WEINLAND, has shown that through contact of the milk-sugar with the intestinal mucosa a substance is produced which is absorbed into the blood and which when it is carried to the pancreas brings about in this organ an increased production of lactase. This phenomenon belongs, therefore, in the same group with those which we discussed under the heading of the secretins.[2] We have here another example of the chemical connection which exists between different organs.

---

[1] VERNON: Journal of Physiology, 1905.   [2] See pp. 211 and 227.

CHAPTER XIII.

THE REGULATION OF THE BILIARY AND INTESTINAL SECRETIONS—THE FUNCTIONS OF THE BILE.

1. The Secretion of Bile.—In discussing the question of the secretion of bile we must distinguish between the formation of the bile in the liver, together with its collection in the gall-bladder, and the flow of the bile into the intestinal canal. The two processes go on independently of each other, and the older experiments made by producing a fistula of the gall-bladder give us an insight only into the first of these questions. What determines the flow of bile into the intestinal canal?

This question has recently been investigated by BRUNO and KLADNIZKI,[1] who have succeeded in turning the orifice of the gall-duct outwards in such a way that the bile flows out upon the skin. This they accomplish by cutting out of the intestine the orifice of the duct with its surrounding mucous membrane, and suturing the latter to the serous coat of the duodenum. The entire loop of intestine is then sewed into the abdominal wound.

Contrary to the older observations made on gall-bladder fistulæ it has been found that the bile does not flow into the intestine all the time. The secretion of bile is intimately connected with taking food and does not begin until a definite time has elapsed after partaking of a meal. The length of this latent period differs with the different kinds of food. The secretion continues as long as the period of digestion,

---

[1] BRUNO and KLADNIZKI: PAWLOW's Work of the Digestive Glands. Translated by THOMPSON. London, 1902, p. 155. BRUNO: Review in Jahresber. d. Thierchemie XXVII, p. 441.

, but not at an even rate, fluctuating in quantity with variations in the nature of the food.

When different food substances are fed separately to a dog operated upon as indicated above, it is found that neither water, acids, raw white of egg, nor boiled starch as a thick or a thin paste causes a flow of bile. Fat, however, and to a less extent extractives of meat, and the digestion products of white of egg set up a free discharge of the secretion. The bile, therefore, is not unlike the gastric or pancreatic juice, which has each its specific excitants.

How now do the various substances which bring about a secretion of bile into the intestine do this? The explanation which has been and is given ordinarily is that the bile flows in consequence of a nervous reflex which is initiated through the action of certain constituents of the intestinal contents upon the mucous membrane of the duodenum. It is much more probable, however, that the connection between duodenum and liver is of a chemical character. This is shown by the experiments of BAYLISS and STARLING, who find that pancreatic secretin not only augments the flow of pancreatic juice but also the discharge of bile into the intestine. It is an interesting fact that the pancreas and liver should have such a common excitant, for as will be shown immediately the bile augments in a most marked way the digestive properties of the pancreatic juice.

2. **The physiological importance of the bile** is shown very clearly in the classical observation of CLAUDE BERNARD and the more modern one of DASTRE.[1] CLAUDE BERNARD noticed that in rabbits, in which the opening of the pancreatic duct into the intestine lies some 30 cm. below that of the bile-duct, the absorption of fat after a fatty meal does not begin until the food has passed the pancreatic duct, for while the lymph leaving the intestine below this point is milky in appearance that above it is clear. This shows that the bile

---

[1] DASTRE: Arch. de phys. norm. et path., 1890, XXII, p. 315.

alone is unable to bring about an absorption of fat. DASTRE'S experiments consisted in producing artificially in dogs the reverse of the above conditions. After ligating the bile-duct he established a fistulous opening between the gall-bladder and the middle of the small intestine. In spite of the fact that the pancreatic juice entered the duodenum as usual, no fat was absorbed from the entire upper half of the small intestine. Not until the biliary fistula was passed did the lacteals show a milky appearance to indicate fat absorption. This shows how important is the function of the bile for the rapid and proper absorption of fats, and indicates clearly that physiological ends are best served when pancreatic juice and bile act together upon fat.

The important rôle of the bile in the digestion of fats can also be shown very clearly in experiments carried out in glass. HEIDENHAIN, WILLIAMS, and MARTIN have all shown that the presence of bile in a reaction mixture of fat and pancreatic lipase markedly increases the velocity of the formation of fatty acid and alcohol. Especially commendable are the recent experiments of RACHFORD, BRUNO, and GLAESSNER, carried out with purified ferments instead of ordinary extracts of the pancreas, or the pure pancreatic and biliary secretions themselves. GLAESSNER'S [1] results are given in the following table and deal with human pancreatic juice. It is interesting to note that what SCHEPOWALNIKOW observed in dogs is true here also, namely, the simultaneous presence of both bile and intestinal juice favors the lipolytic activity of the pancreatic juice more than that of either one by itself. The last column indicates the percent of fatty acid formed in twenty-four hours.

	Fatty acid.
100 c.c. olive-oil + 20 c.c. pancreatic juice.	$=22\%$
100 c.c. olive-oil + 20 c.c. pancreatic juice + 10 c.c. bile.	$=30\%$
100 c.c. olive-oil + 20 c.c. pancreatic juice + 10 c.c. intestinal juice	$=35\%$
100 c.c. olive-oil + 20 c.c. pancreatic juice + 10 c.c. bile + 10 c.c. intestinal juice.	$=40\%$

[1] GLAESSNER: Zeitschr. f. physiol. Chem., XL, p. 470.

A long-accepted explanation of the rôle of bile in hastening fat digestion has been sought in its power of favoring the emulsification of fats. Through emulsification of the fat the amount of surface exposed to the action of lipase is greatly increased. Experiments carried out by HEWLETT [1] indicate that bile acts in yet another and even more direct way than this in favoring the activity of the ferment. These experiments show at the same time which constituent of the bile it is that favors the fat-splitting activity of the pancreatic juice, for bile is a mixture of a number of chemical entities. ✗

If, instead of the ordinary very sparingly soluble fats, the soluble triacetin is used, so that the question of emulsification plays no part whatsoever, it is found that bile still markedly hastens the action of pancreatic juice upon this substance. In one experiment, for example, pure pancreatic juice obtained from a dog by secretin and atropin injections produced in one hour at 20° C. enough acid to require 0.5 c.c. 1/20 normal alkali solution to neutralize it, and in twenty-four hours 12.6 c.c. In another tube which contained the same amount of pancreatic juice and triacetin, but in addition a small amount of bile, the acidity produced in one hour amounted to 13.0 c.c., in twenty-four hours to 18.6 c.c. of a 1/20 normal alkali solution. That the acceleration of the decomposition in the later hours of the experiment should be less than in the earlier is readily explained by the fact that the reaction is approaching an equilibrium.

If now an attempt is made to discover which constituent of the bile it is that favors in this way the decomposition of triacetin under the influence of pancreatic juice, it is found, first of all, that boiling the bile does not destroy this property. It is therefore probable that we are not dealing with the action of any enzyme contained in the bile. HEWLETT has further shown that this property resides neither in the cholesterin nor in the bile pigments, nor in variations in the reac-

---

[1] HEWLETT: Johns Hopkins Hospital Bulletin, 1905, XVI, p. 20.

tion or in the amount of calcium salts present. All the accelerating effects of bile upon the lipolytic activity of pancreatic juice can be equally well produced by the addition of lecithin. This is shown very well in the following table, in which is indicated the number of cubic centimeters of a 1/20 normal alkali solution which were required to neutralize the acid formed in twenty-four hours in three tubes containing the same amounts of pancreatic juice and triacetin.

Pancreatic juice + triacetin .... = 4.3 c.c.
Pancreatic juice + triacetin + 2 c.c. bile .... = 19.5 c.c.
Pancreatic juice + triacetin + 2 drops alcoholic solution of
lecithin....................................... = 19.9 c.c.

Commercial bile salts also accelerate the action of pancreatic juice on triacetin, but when these salts are purified they lose this power, which seems to indicate that the action of the crude preparation is determined solely by its contamination with lecithin.

As to the manner in which the bile assists the lipolytic action of the pancreatic juice, we have as yet no satisfactory explanation.

Bile also favors the action of some of the other ferments contained in the pancreatic juice. But while bile may even treble the velocity with which fats are digested, it only doubles the velocity with which amylase will act on starch, or alkali-proteinase (trypsin) on proteins.

Aside from its action as an aid to the pancreatic ferments, bile has yet other, though perhaps scarcely as important, functions. While a large number of the fatty acids are freely soluble in water, this is not true of the fatty acids derived from most of the fats (stearin, palmitin, olein) which constitute our food. MOORE, ROCKWOOD, and PFLÜGER's [1] studies

---

[1] MOORE and ROCKWOOD: Journal of Physiology, 1897, XXI, p. 58.
PFLÜGER's numerous papers on fat are contained in Pflüger's Archiv, Vols. LXXX to LXXXVI (1900 to 1901).

are therefore of great importance, which show that a mixture of bile and sodium carbonate is able to dissolve large amounts of stearic and palmitic acids. Bile is able to keep even the insoluble calcium and magnesium soaps in solution.

The presence of bile in a reaction mixture of protein and gastric juice retards the action of the ferment greatly. It seems plausible, therefore, that the bile is of physiological importance by interfering with the activity of the gastric juice after this escapes into the duodenum from the stomach.

To the bile has also been attributed an antiseptic action, and it has been believed that through its presence the development of bacteria throughout the alimentary- tract is markedly inhibited. Careful study seems to indicate, however, that this antiseptic action is only very weak, if it is present at all. Bacteria develop freely in bile itself and culture media containing bile. The increased putrefaction of the alimentary contents, which is observed in at least some cases of icterus, must therefore be attributed to other causes. Foremost among these must stand the less perfect absorption of the foodstuffs whose presence in the alimentary tract furnishes a ready culture ground for the various bacteria found here.

The bile is believed to aid intestinal peristalsis. This idea seems borne out by clinical observation, though laboratory experiments in this direction have brought no unequivocal results. If it is true that a lack of bile leads to a decreased peristalsis, we could readily find in the retention of alimentary contents from this cause an additional reason for the increased intestinal putrefaction found in these cases.

We must, in conclusion, call attention to a function of the bile which is still questioned by some authors. According to some recent experiments, it is claimed that lipase is secreted in the pancreatic juice only in an inactive form, that is, as a proferment or zymogen, and that this inactive form is activated through the bile. The bile would therefore serve the same important function in fat digestion that the intestinal juice serves through the enterokinase it contains in protein digestion.

3. **Regulation of the Intestinal Secretion.**—In order to obtain a collection of pure intestinal juice use is made of THIRY's classical fistula, or THIRY's fistula as modified by VELLA. An opening is made in the abdominal wall of an animal, and a suitable loop of intestine is pulled out. Two transverse cuts separate any desired length of the intestine from the main portion of the tube, the continuity of which is restored by an end-to-end anastomosis. The separated loop of intestine is then closed at one end by a purse-string suture, while the opposite end is sewed into the edges of the wound. In this way a test-tube shaped piece of the bowel is separated from the main body of the gut (THIRY fistula), or both ends may be sewed into the abdominal wound when we have the so-called THIRY-VELLA fistula. The artificial production of all the most successful fistulæ along the gastro-intestinal tract is modelled after the original THIRY operation on the small intestine.

In the quiescent state of the animal practically no secretion can be obtained from such an isolated loop of intestine. As soon as food, such as starch paste, sugar, or peptone, is introduced into the loop an increased secretion takes place. Apparently, therefore, intestinal juice is secreted under ordinary circumstances only by those portions of the gut with which food is in contact, and not throughout its entire length. The presence of food may cause a secretion either through its mechanical or chemical properties. That mere mechanical stimulation may be effective in causing a secretion of intestinal juice is proved by the fact that the irritation of a cannula in an intestinal fistula, a rubber ball, etc., all lead to a secretion of juice. According to PAWLOW and his coworkers the intestinal juice obtained in consequence of mechanical stimulation alone differs from that obtained by chemical means. The former kind consists chiefly of water and salts with but very little or no *enterokinase.*[1]   When the secretion is poured out upon food the intestinal juice is rich in enterokinase.   Apparently much more powerful than the food itself in bringing about an intestinal secretion are the pancreatic ferments which, under ordinary circumstances, accompany the food along the intestine.   An isolated loop which is secreting little or no juice will become active very soon after a few cubic centimeters of pancreatic juice are put into it.   If the pancreatic juice is previously boiled, this property is lost.   Which of the pancreatic ferments is active in this direction is not yet known.

Observers agree that the normal enteric juice is a thin, slightly yellowish liquid, ordinarily said to be alkaline in reaction.   The fluid which collects in the course of a few weeks in loops of intestine which are ligatured off from the main tube probably does not represent a normal secretion. Usually this is more or less gelatinous and represents no doubt the inspissated intestinal juice plus the mucin and other abnormal substances poured out by the loop in consequence of the "catarrh" which arises in it.

The nervous system no doubt influences the secretion from the small intestine, but in just what way is not known. The statements made by different observers contradict each other in many points.   A secretion of fluid occurs into the intestine when all the nerves passing to the loop have been severed.   This so-called "paralytic secretion" contains, according to MENDEL's observations, all the ferments present in the normal juice.   A secretion of fluid into isolated loops of intestine when entirely removed from the body also occurs. All the structures necessary for secretion must therefore be contained in the wall of the gut, and it would no doubt be straining a point should we attribute this secretion to the nervous plexuses found in the wall of the intestine, and not solely to the mucous membrane.   The fluid secreted into loops entirely removed from the body is not to be regarded

---

[1] See p. 246.

as normal intestinal juice. For this we need the coöperation of the circulating blood, from which are taken the substances which either directly or after they have been changed into new chemical compounds through the activities of the mucous membrane go to make up the intestinal juice.

Stimulation of the vagus nerve in the neck or below the diaphragm seems not to affect the secretion of enteric juice. Removal of the cœliac or mesenteric plexuses seems to bring about an increased secretion similar, perhaps, to the ordinary paralytic secretion but not so great in amount. If certain of the ganglia are left behind, the secretion may not take place.

Attention has already been called to the fact that the intestinal juice has not the same chemical composition throughout its entire length. Special mention must be made of the secretion of the duodenum. In addition to the glands of LIEBERKÜHN found throughout the whole of the small intestine the duodenum contains the glands of BRUNNER. Histologically these resemble the glands of the stomach. The debate which has long been carried on as to whether the proteolytic ferment found in the duodenal juice is acid- or alkali-proteinase, or amphoproteinase can be looked upon as decided through ABDERHALDEN's work. While it is impossible under certain circumstances to say whether we are dealing with pepsin or trypsin when a fluid containing one or both of them is allowed to act on fibrin or any other complex protein, it is possible to do this when polypeptides are used. Alkali-proteinase will split certain polypeptides not affected by acid-proteinase and *vice versa.* In this way it has been possible to prove that the proteolytic powers of the duodenal juice are due to acid-proteinase (pepsin) secreted by the glands of BRUNNER. As is universally the case, the proteolytic ferment of the duodenal juice is also accompanied by a milk-curdling ferment. As a whole, therefore, the duodenal juice is very like gastric juice, yet the physiological importance of the latter when compared with the former is not great.

The secretion of the *large intestine* is small in amount and rich in mucin. It is stated to be alkaline in reaction, and to contain no ferments in sufficient amount to be of physiological importance. The large intestine acts chiefly as an organ of absorption.

4. Enterokinase.—This is the term given by PAWLOW to a substance discovered by SCHEPOWALNIKOW in the mucous membrane of the small intestine which has the interesting property of inaugurating or at least of increasing enormously the proteolytic power of the pancreatic juice as it flows from the gland.

We are probably correct in believing that neither the pancreas nor the pancreatic juice obtained directly from the pancreatic duct contains any alkali-proteinase (trypsin), but only a substance (the so-called proferment or zymogen) which can be converted into this ferment through contact with enterokinase. Under physiological conditions this contact with enterokinase is established as soon as the pancreatic juice flows over the mucous membrane of the small intestine, the walls and secretions of which contain this activating substance. The amount of mucous membrane necessary to bring about at least some activation of the otherwise inactive pancreatic juice is very small according to DELEZENNE and FROUIN'S [1] experiments. In the ordinary method of making a pancreatic fistula according to the method of HEIDENHAIN and PAWLOW,[2] by cutting out a small portion of the mucous membrane surrounding the pancreatic duct and sewing this into the edge of the abdominal wound, the escape of the pancreatic secretion over this bit of mucous membrane is sufficient to give it well-marked digestive properties for proteins. If the escape of the juice across the transplanted mucous membrane is avoided by inserting a cannula into the duct above

---

[1] DELEZENNE and FROUIN: Compt. rend. de Soc. biol., 1902 CXXXIV, p. 1524.

[2] See p. 215.

it, juice entirely incapable of digesting egg albumin is obtained. This juice readily becomes active in this direction if a small amount (a drop or two) of intestinal juice obtained from an intestinal fistula in another animal is added to it.

As to the *nature* of enterokinase but little is known at present, for the substance cannot be obtained in even an approximately pure state. Since it is readily destroyed at comparatively low temperatures PAWLOW has looked upon it as a ferment. This conception of enterokinase scarcely harmonizes with the observations that have been made, which indicate that a definite amount of enterokinase-containing fluid can activate only a limited amount of pancreatic juice. True ferments, it is well-known, act upon an infinite amount of substance if only sufficient time is given.

The secretion of enterokinase is not to be looked upon as a function performed by the small intestine at all times. Enterokinase appears and disappears from the juice poured out by the upper portions of the small intestine in the same way as the amount of bile poured into the duodenum is controlled by such circumstances at the taking of food. In fact, the secretion of enterokinase is connected with physiological processes going on in the intestine in the same way as the secretion of bile. When the small intestine is stimulated mechanically it secretes a juice, but it is thin and watery and contains practically no enterokinase. As soon, however, as a few cubic centimeters of pancreatic juice are introduced into the lumen of the intestine, the juice secreted becomes rich in enterokinase. Boiled pancreatic juice does not have this power. The secretion of the watery constituents and of the enterokinase of the intestinal juice represent, therefore, different physiological processes.

The idea that pancreatic juice obtained directly from the pancreatic duct has no power to digest proteins contradicts the views of a number of observers who have claimed that *the spleen* furnishes at the height of digestion a substance which is absorbed into the blood and through its action on the pancreas changes the proferment found here into alkali-proteinase. This function of the spleen we may no doubt now say does not exist, and may safely attribute the results obtained by the workers on the spleen to accidental bacterial contamination of the extracts with which they worked. We know now that bacteria can also activate the proferment of alkali-proteinase, and presume that they are able to do this because of a "kinase" which they also contain.

Attention has several times been called to the physiological connection which exists between different portions of the alimentary tract. It may not be amiss to do it once more at this place. The bile and the intestinal juice, which so markedly increase the activities of the ferments found in the pancreatic secretion are poured into the intestine in an apparently entirely purposeful manner. BAYLISS and STARLING have found that pancreatic secretin, which so markedly increases the discharge of pancreatic juice, increases also the flow of bile. If a cannula is tied into the bile-duct after previous ligature of the gall-bladder, the intravenous injection of secretin brings about not only a flow of pancreatic juice but also an augmented flow of bile. According to DE-LEZENNE pancreatic secretin increases also a discharge of intestinal juice containing enterokinase from the upper portions of the small intestine. Through such means the combined action of pancreatic juice, intestinal juice, and bile upon the food as it escapes from the stomach is secured; in other words, the conditions are established which experiment has shown to be the most favorable for the rapid digestion (and absorption) of the various foodstuffs.

CHAPTER XIV.

THE ALIMENTARY TRACT AS AN ABSORPTIVE SYSTEM.

1. The Problem of Absorption.—The problem of the absorption of foodstuffs from the alimentary tract is the same as the problem of absorption in general. It is self-evident that absorption and secretion are really only different phases of the same thing, for in the one case we are asked to explain why a tissue takes up a certain chemical substance, while in the other we are called upon to tell how a tissue rids itself of this same chemical substance. In the last anlaysis we have to answer these questions for every individual cell, for each absorbs certain substances and secretes others. Very often one and the same cell absorbs and secretes the same substance. As will become apparent later the intestinal epithelium, for instance, absorbs fat from the lumen of the gut and secretes it into the lymph stream. The lymph may therefore be looked upon as an absorptive (fluid) tissue which in turn becomes a secretory tissue when the body cells begin to take the fat away from it.

Put briefly, therefore, we can say that in considering the problem of absorption we have to answer the question, How do the various chemical substances pass from one cell into another, or from one cell into a liquid (such as the lymph or blood), or finally from such a liquid into a cell? Under the last heading comes, for example, the passage of the chemical substances contained in the lumen of the alimentary tract into the cells lining this tract, while the passage of these same substances into the lymph or blood stream illustrates the second of the above headings.

The natures of the chemical substances which pass in this way from one cell to another or from one tissue into another of necessity differ greatly from each other. Limiting ourselves for the moment simply to the food ingested at an ordinary mixed meal we can readily recognize the very large number of different chemical compounds with which we have to deal. It is possible to group all of them chemically under a few headings,—the proteins, the carbohydrates, the fats, the salts, and water. As the problem of absorption is a physico-chemical one we will regroup them in a somewhat different way as the colloids, the crystalloids, and water.

Still more difficult is it to say what forces are active in bringing about the movement of these various chemical substances. One of the most readily intelligible of these is, perhaps, *diffusion*, which is nothing but an expression of differences in osmotic pressure. *Capillary forces* must also be considered in a discussion of the means by which chemical substances move from place to place. Finally we can mention the variable *mechanical affinity* (OSTWALD) of colloids for water and substances dissolved in the water. The forces active here are not as yet clearly understood. They may be capillary in character or connected with the enormous surface presented by colloids. At any rate, modern investigations support strongly the idea that an understanding of the laws underlying the absorption of water by colloids will do more to give us an insight into the phenomena of absorption and secretion than the laws of osmotic pressure and diffusion have ever done.

But the problem of absorption is not stated when we say that we have to explain the movement of a large number of different chemical substances under the influence of different forces. The activity of these forces is largely altered by the fact that the passage of the different chemical substances from cell to cell, or from liquid to cell, or from cell back to liquid, is modified by the existence of differences in the permeability of protoplasm, more particularly by the existence of cell membranes which are sometimes permeable, sometimes impermeable, or again partially permeable to a diffusing substance. Not only do different cells differ in their permeability to certain chemical substances, but the same cell may under different conditions be at times permeable, at others impermeable, to the same substance.

When we consider that under ordinary circumstances the different chemical substances, the different forces, and the different permeabilities of protoplasm are all working together in making up a picture of absorption as we witness it in a physiological experiment some idea may be obtained of the difficulties which face the investigator who attempts the solution of the problem. Nevertheless great strides have been made within recent years in substituting known laws of physics and chemistry for the vitalistic explanations of the older observers. In the following paragraphs are discussed in brief the physical chemistry of the substances which serve as food, the forces active in bringing about their absorption, and the membranes which alter so markedly the independent activities of the other two. It is beyond the scope of this volume to enter more deeply than this into the subject.

2. **The Physical Character of the Foodstuffs. Colloids and Crystalloids.**—From the standpoint of absorption the chemical constitution of the foods which we consume plays less of a rôle than their physical character. It is for this reason that a regrouping of these substances into colloids, crystalloids, and water has been suggested, for, as we shall see, the readiness with which they diffuse, for example, is of greater import than the arrangement of the atoms which constitute their molecules.

As many as fifty years ago GRAHAM recognized that different chemical substances differ greatly in the rate with which they diffuse through solvents of various kinds. Those which diffuse very slowly are for the most part amorphous, and since ordinary glue is an example of such he called the substances belonging to this class *colloids.* On the other hand, those which diffuse rapidly he called *crystalloids,* for the bodies belonging in this class are mostly crystalline, as sugar and salt. From the physiological standpoint of absorption this difference in the rates of diffusion still stands as one of the most important differences between these two classes of compounds, for, as we shall see, our food is made up to a large extent of typical representatives of these classes.

Modern advances in physical chemistry have given us other criteria besides differences in the rates of diffusion and in their amorphous or crystalline constitution by which we can distinguish between colloids and crystalloids. When dissolved in water or other solvents the colloids do not form true solutions, but remain suspended in the liquid. Colloidal solutions are, therefore, heterogeneous. More correctly put, a colloidal solution represents a mixture of two substances which are only partially soluble in each other. We formerly looked upon crystalloidal solutions as homogeneous. Recent experiments indicate, however, that these too are heterogeneous, only much less markedly so than the colloidal solutions.

Solutions of crystalloids show an osmotic pressure which is proportional to the number of particles of dissolved substance contained in the unit volume of the solvent. As will be seen later it is upon this fact as well as upon the minuteness of the dissolved particles that the great diffusibility of the crystalloids depends. In contrast herewith the so-called "typical" colloids show no osmotic pressure and in consequence do not diffuse at all. But only very few "typical" colloids exist; the vast majority show some osmotic pressure and some diffusibility, even though it be but slight.

These enormous differences in osmotic pressure between crystalloids and colloids correspond to similar differences in the molecular weight of the substances composing the two classes. While the molecular weight of most crystalloids is relatively low, that of the colloids is very high. The molecular weight of glue is, for example, about 6000, that of colloidal tungstic acid 1700.

The following table[1] shows most clearly how a high molecular weight and osmotic activity are antagonistic values. The figures refer to 10 percent solutions of the various substances.

Substance.	Mol. Wt.	Osmotic Pressure in Atmospheres.	Depression of Freezing-point.
Methyl alcohol.............	32	70.00	5.781
Urea.....................	60	37.34	3.084
Glucose..................	180	12.43	1.027
Cane-sugar	342	6.54	0.540
Albumose	2400	0.93	0.078
Albumin	13000	0.17	0.015

Crystalloids can, moreover, diffuse uninterruptedly through colloidal membranes, such as animal bladders, intestines, sheets of agar-agar, or gelatine, etc. Colloids are for the most part unable to do this. Upon this fact is based the principle of dialysis, in which crystalloids are separated from colloids by placing the mixture in a tube of parchment or an animal bladder, and hanging the whole in water or some other solvent. The crystalloids diffuse out, leaving the colloids behind.

It must be pointed out at once, however, that no sharp line exists between the colloids, on the one hand, and the crystalloids, on the other. Between the two extremes representing the typical members of these groups there are found an infinite number of substances, which lean more or less strongly toward one side or the other. To illustrate this fact we need only mention that not every crystalloid diffuses through animal membranes, and not every colloid

---

[1] Höber: Zelle und Gewebe. Leipzig, 1902, p. 19.

is incapable of doing so. Moreover, certain colloids may be obtained artificially, not only in an amorphous state but also in a most beautifully crystalline form (HOFMEISTER).

So far as absorption is concerned, we can say that nearly every food which enters the alimentary tract belongs in one or the other of these two groups, or does so after it is acted upon by the digestive juices. As examples of the colloids we may mention all the different albumins, globulins, albuminoids, and starch paste; under the crystalloids, the various sugars, salts, acids, and alkalies. We begin to see now the importance of the various digestive processes which go on in the intestine. While it will be shown later that albumins, for instance, may perhaps be absorbed as such, possibly are even absorbed in part in an unaltered state, it is self evident that as these colloids become more like crystalloids their diffusibility, and in consequence their absorption, will be greatly facilitated. In the process of digestion this transference from the side of the non-diffusible colloids to that of the diffusible crystalloids does in fact occur. The peptones are much less colloidal in character than the albumins from which they come, and the ultimate products of digestion formed under the influence of the proteinases or protease are practically all typical crystalloids. Starch paste also leaves the side of the colloids when acted upon by amylase and as maltose becomes grouped with the crystalloids. The fats which as such are incapable of diffusion diffuse readily after having been acted upon by lipase and changed into fatty acid and glycerine.

3. Membranes.—We shall consider next not the forces that bring about absorption, which would seem most logical, but rather the obstacles which modify absorption—namely, membranes of all kinds. This will make what is to follow more intelligible.

As we are interested in membranes chiefly from the stan point of whether they allow substances to diffuse through them or not and to what extent they permit this, the classifi-

cation into *semipermeable* and *permeable* membranes is probably best suited for our purposes.

A true *semipermeable membrane* is one which allows only
the solvent and none of the substances dissolved in the solvent
to pass through. True semipermeable membranes are really
very rare. The existence of such membranes was discovered
by TRAUBE, but we are indebted to PFEFFER for the idea of
supporting these in unglazed vessels of earthenware, so that
accurate studies of them could be made. The nature of true
semipermeable membranes may be made somewhat clearer
if the method of making a so-called *precipitation membrane*
is described.

An ordinary PASTEUR-CHAMBERLAND filter is sawed in half
transversely (see Fig. 28, p. 263). The resulting small clay
cylinder after proper washing is filled with a solution of
copper sulphate, and the whole is then dipped into a second
vessel containing a potassium ferrocyanide solution. The two
solutions penetrate the unglazed clay wall from opposite
sides, meeting in the middle, where a precipitate of copper
ferrocyanide is produced.

This precipitate of copper ferrocyanide constitutes a semipermeable membrane, that is, one which is permeable to
water but not to substances dissolved in the water. We
must point out at once, however, that this statement is not
strictly true. It is true for the salts which have been used to
produce the precipitation membrane, but a certain amount
of nearly all other substances can pass through such a membrane. Strictly speaking, therefore, even the semipermeable membranes are permeable to some extent, though, as
we shall see, not at all to the same degree as the truly permeable membranes. Other precipitates have also been used as
semipermeable membranes, such as zinc ferrocyanide and
calcium phosphate, but copper ferrocyanide is perhaps the
best.

Almost any one of the animal membranes, such as the
bladder of various animals, portions of intestine, or ordinary

parchment paper, may be taken as a type of a *permeable membrane.* By this is meant that these allow most crystalloids which are held back by semipermeable membranes to pass through them with ease. But substances having a high molecular weight (as the various colloids) pass through these membranes not at all or only to a slight extent.

Do true semipermeable membranes exist in the animal organism? So far as we know they do not. Many cells are impermeable to a large number of substances, but all are permeable to some dissolved substances, The red blood-corpuscles, for example, are impermeable to many different salts, but they readily allow ammonium compounds, urea, etc., to pass into them. The majority of cells are even more permeable than the red blood-corpuscles, and this even to very large molecules. The fact that colloids can to some extent diffuse into other colloids already points in this dire tion, and later we shall become acquainted with experiments in which colloidal sodium silicate was found to be absorbed from the intestinal tract and excreted in the urine. The intestinal epithelium must therefore be looked upon as made up of cells which are permeable to at least certain very large molecular aggregates, if only sufficient time be allowed for their absorption. But the intestinal epithelium is not equally permeable to all substances, even when they possess approximately the same molecular weight, and show in pure solvents approximately the same rates of diffusion. What is true for the intestinal epithelium holds still more when we deal with different tissues. Each tissue varies in the ease with which it will absorb different chemical compounds. This selective permeability is probably more confusing in endeavoring to unravel the problem of absorption than any one other factor, though, as will be seen shortly, great strides have been made in the solution of this problem within the last decade.

We have not as yet said where these different membrane exist in tissues. Every cell is surrounded by a membrane but the *physiological membrane* with which we are dealing here need not, and in fact usually does not, coincide with the *morphological cell-wall.* This is shown particularly well in vegetable cells in which the physiological membrane lies entirely within the morphological membrane. Certain cells do not have any morphological membrane at all, yet a physiological membrane permeable to certain substances and impermeable to others no doubt exists in these cases. This is so, for example, in the red blood-corpuscles, in amœbæ, and in the intestinal epithelium. Finally, when we deal with tissues, whole cells arranged in layers may constitute a membrane through which diffusion occurs from one medium into another. Under these circumstances the protoplasm of the cell, as a whole, constitutes the membrane.

What has been said of the cells themselves holds true also for groupings of these cells as they exist in the various membranes of the body—such as the absorptive mucous membranes of the alimentary and urinary tracts, the synovial membranes, etc. As the individual cells differ from each other in permeability, so do also the various tissues built up of these cells. In dealing with tissues we have yet to take into consideration the permeability of the intercellular substance. This may be entirely different from the permeability of the cells themselves.

4. The Forces Active in Absorption.—If an odorous gas is liberated in one corner of a room, it soon spreads throughout the whole room, so that it may be detected anywhere in it. We say that the gas *diffuses* through the room, and, according to the kinetic theory, this diffusion takes place because the molecules of the gas are in constant motion and move in all directions. The odorous gas continues to diffuse, if nothing prevents it, until the concentration of this substance is the same in all portions of the room.

If a substance (such a copper sulphate) is put into a vessel and a solvent (such as distilled water) is carefully poured upon it, we find that after a while the soluble substance has distributed itself uniformly throughout the solution. We say that the soluble substance has *diffused* through the solvent. This process of diffusion is analogous to the diffusion of gases described in the preceding paragraph. When a crystal of copper sulphate is covered with pure water, the diffusion of the copper salt shows itself as a blue zone about the crystal, which gradually spreads until all the water is tinged uniformly blue and the crystal (provided it has not been too large) has entirely disappeared. It is not necessary to start with a crystal of the copper sulphate,—a solution of this salt may be used instead and it be carefully covered with distilled water. Diffusion occurs in the same way and continues until the entire volume of water is tinged uniformly blue. What has been said holds for any solvent and any soluble substance, be the latter a solid, a liquid, or a gas.

What is the cause of this movement of particles of dissolved substance from places of higher concentration to places of lower concentration, in other words, this *diffusion?*

In the case of the odorous gas liberated in a room, we attribute the spread of the odorous substance throughout the room to the fact that its partial pressure is greater at the point of liberation than anywhere else in the room. In other words, the gas pressure is highest where the odorous substance is liberated, and in consequence the odorous particles of gas are driven through the room until the pressure is everywhere the same. Entirely analogous to the movement of the particles of a gas through a vacuum or another gas is the movement of the particles of a dissolved substance through a solvent, only, while we call the force which causes the movements of a gas, gas pressure, we call the force which causes the movement of a dissolved substance *osmotic pressure.* Just as a gas exerts a certain (gas) pressure upon the walls of its container, so a dissolved substance exerts a certain (osmotic) pressure upon the walls of its container. This pressure, indicative of the movement of the dissolved substance, we can render apparent by separating the region of higher concentration from that of lower concentration by a *semipermeable membrane.* As was pointed out before, this means a membrane which will not give passage to a dissolved substance but only to its solvent. The dissolved substance in its movement through the solvent is then stopped by this wall, in consequence of which it exerts a pressure (osmotic pressure) upon it which evidences itself, if the membrane is not so supported as to prevent it, by a bulging of the membrane toward the region of lower osmotic pressure.

Just as gas pressure varies with different external conditions, so does osmotic pressure (VAN'T HOFF). Of great physiological importance is the fact that at constant temperature *the osmotic pressure of dilute solutions is proportional to the concentration of the dissolved substance.* By the concentration of the dissolved substance is meant the number of particles of this substance contained in the unit volume. This, the first law of VAN'T HOFF, means that if a solution of a certain concentration has a certain osmotic pressure, the osmotic pressure will be doubled if in the same volume of solvent twice the amount of soluble substance be dissolved, or trebled if three times the original amount goes into solution in it (see below).

VAN'T HOFF's second law is also of physiological importance. *The osmotic pressure of a dilute solution is proportional to the absolute temperature,* in other words, increases as the temperature is increased, even when all other external conditions, are left unchanged. The third law of VAN'T HOFF, that *at the same temperature equal volumes of all dilute solutions which have the same osmotic pressure contain the same number of molecules* does not interest us in a discussion of the immediate problem.

We have purposely spoken above of the osmotic pressure of dissolved *particles.* We can say now that wherever this term has been used above we may substitute the word *molecules,* but this only when the substance under consideration is a so-called *non-electrolyte*, that is, a substance which in solution in water does not conduct the electric current. Into this group of non-electrolytes belong, for example, the various sugars, glycerine, and urea. Under the *electrolytes* are classed all those substances which when dissolved in water conduct the electric current. This group is composed of the *acids*, *bases*, and *salts*, the more typical examples being the *strong* acids, bases, and salts, such as the mineral acids, the caustic alkalies, and the salts formed by the chemical union of these two.

Van't Hoff's laws do not hold in this unmodified way for the electrolytes. Solutions of electrolytes all behave as though they contained a larger number of molecules than is indicated by the weight of the substance dissolved in the water. This apparent exception to the laws of van't Hoff has been explained by Arrhenius, who has shown that when electrolytes are dissolved in water the molecules break up into smaller electrically charged atoms or groups of atoms called *ions*. This theory of ions is also known as the theory of *electrolytic dissociation*. The degree of dissociation or ionization which the molecules of an electrolyte undergo varies not only with the nature of the electrolyte itself, but also with a number of external conditions such as temperature, concentration, etc. But the amount of this dissociation can be determined experimentally, and when it is taken into consideration the laws of van't Hoff are valid for solutions of electrolytes also. Modified in this way the first law of van't Hoff reads: The osmotic pressure of a solution is proportional to the number of molecules plus ions present in the unit volume of solvent. The word *particles* was used above to cover this conception of molecules plus ions.

What has been said will serve to indicate the important rôle which diffusion must play in determining the migration of dissolved substances from regions of higher concentration to those of lower. Because of diffusion the sugars, the salts, and the digestion products of the proteins and fats enter the intestinal mucosa and pass into the blood and lymph streams. The process of diffusion within the animal organism does not go on as simply and as uninterruptedly, however, as in a vessel containing a solvent and a soluble substance. The rate of entrance of even the simplest chemical substances into the epithelial cells of the intestinal mucosa is far different from the rates of diffusion of these same substances outside of the body. As will become apparent later an important reason for this is to be found in the constitution of protoplasm itself, and of the membranes surrounding this protoplasm.

CHAPTER XV.

THE ALIMENTARY TRACT AS AN ABSORPTIVE SYSTEM
(Continued).

5. The Absorption of Water.—What has gone before indicates how osmotic pressure is one of the forces which determines the movement of dissolved substances.—It will be shown now that under proper conditions it becomes a force which determines the movement of water, and that it is in part responsible for the absorption of this substance in the living organism. The amount of water that can be absorbed from the alimentary tract is enormous. It is ordinarily stated that 4 or 5 liters of pure water may be absorbed in the course of a day. This does not, however, indicate all that can be absorbed. FRIEDRICH MÜLLER once showed in his clinic in Munich a man who not infrequently consumed between 20 and 30 liters of beer in twenty-four hours without having a diarrhœa. Beer does not, of course, represent pure water, and the limits for this substance may be lower.

Osmotic pressure becomes effective in determining the absorption of water when the diffusion of the dissolved particles to regions of a lower concentration is prevented by a semi-permeable membrane. This may be illustrated by the accompanying diagram (Fig. 28). A represents in cross-section a PASTEUR-CHAMBERLAND filter, in the wall of which is deposited a semipermeable precipitation membrane M, such as copper ferrocyanide, made as described above (page 255). The *osmotic cell*, as such an apparatus is called, is closed with a rubber stopper, C, through which passes the glass manometer

262

tube *T*. If this cell is filled with a sugar solution, and the whole is dipped into a second vessel, *B*, filled with water, the sugar endeavors to pass out into the water (diffusion). This movement is prevented, however, by the semipermeable

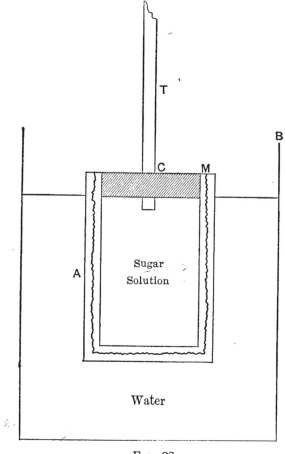

Fig. 28.

membrane and water is in consequence drawn into the osmotic cell, which evidences itself by a rise of the meniscus in the tube *T*. The water continues to enter the cell until the hydrostatic pressure in the tube is equal to the osmotic pressure in the cell.

The reason why the water enters the cell, or, in general,

why water moves from a region of lower concentration to one of higher, is not yet entirely understood. The old idea was that the dissolved particles "attracted" the water. A more correct explanation of the phenomenon based upon differences in surface tension seems to be the following. Liquids (water in this case) are surrounded by a contractile surface film, in consequence of which they always tend to occupy as small a space as possible (that is, tend to become spherical). These surface films, therefore, exert a pressure in a direction toward the centre of the liquid, as shown in Fig. 29, a. The pressure exerted by the diffusion of dissolved particles is evidently opposite in nature to that exerted by

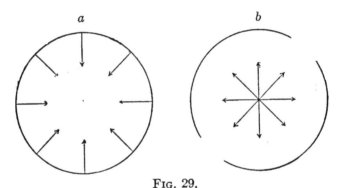

FIG. 29.

such surface films, for the particles of a dissolved substance move from the centre toward the surface of a liquid and press upon it. This is indicated in Fig. 29, b. In the vessel B in Fig. 28, which contains only pure water, we are dealing with surface tension only, which we will represent by P. In the osmotic cell we have this same surface tension, but it is counteracted by the osmotic pressure p. Evidently, therefore, $P > P - p$. Or, to put the same in words, the surface tension of the water outside of the cell is greater than the surface tension inside of the cell minus the osmotic pressure. Water passes from the outer vessel into the cell, therefore, because it is *squeezed* in and not because it is *pulled* in.

It is clear that what has been said regarding pure water

and a sugar solution holds for any pure solvent and any solution. Furthermore, it holds for any two solutions which have not the same osmotic pressure (that is to say, have not the same number of particles dissolved in the unit volume of solvent). The solvent moves always from the region of lower concentration to that of higher concentration (that is, from the region of lower osmotic pressure to the region of higher) when the two are separated by a semipermeable membrane, and this movement continues until the osmotic pressure on both sides is the same. It is clear, therefore, that if the osmotic cell is filled with water instead of with a sugar solution and the surrounding vessel contains a sugar solution, or, to put it more generally, if the osmotic cell contains a solution of a lower concentration than the surrounding vessel, the water will move out of the cell into the solution in the outer vessel until the osmotic pressure on both sides of the membrane is again the same.

Do similar conditions exist in the case of the body cells? To a certain extent they do. It was pointed out above that many cells which have been studied seem to be surrounded by membranes which have to be considered as approxi-

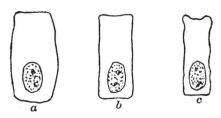

FIG. 30.

mating true semipermeable ones more or less perfectly. Certain cells even seem to possess membranes which are truly semipermeable for a large number of substances. Fig. 30 illustrates the behavior of any cell possessing a truly semipermeable membrane towards solutions of various kinds.

If we consider *b* the cell under normal physiological conditions, that is, as in osmotic equilibrium with the liquids with which it is bathed, then *a* and *c* represent respectively the effect upon this cell of immersion in a solution more dilute or more concentrated than that in which it is immersed normally. If the cell is placed in a solution which has a lower osmotic pressure than the cell contents, then water will pass into the cell until the osmotic pressure on both sides of the (physiologically) semipermeable membrane is the same. The cell will in consequence increase in size (if some external obstacle does not prevent it) and may swell so much that the cell-wall is ruptured and the cell contents are allowed to escape. The reverse must occur when the cell is immersed in a fluid having a higher osmotic pressure than the cell contents. Under these circumstances water passes through the semipermeable membrane surrounding the cell in the direction from within outwards until osmotic equilibrium is once more restored. As the water passes out of the cell, this shrinks, and the amount of the shrinkage is determined by the amount of water given off by the cell. This in turn is determined by the amount of the difference between the osmotic pressure of the cell contents and that of the surrounding liquid.

True semipermeable membranes are rarely found surrounding any cells that have been accurately studied, and so it will not seem strange that physiological experiment has shown that the epithelial cells of the alimentary tract obey the ordinary laws governing the osmotic absorption of water only when exposed to great and sudden changes by being flooded with solutions of very high or very low osmotic pressures. The reason for this is readily intelligible when it is remembered how very permeable the alimentary mucosa must be to allow the passage of the host of soluble substances which daily pass through it into the blood or lymph or from these circulating fluids out into the lumen of the digestive tube. Whenever a membrane allows the passage of a substance dissolved in a liquid found on either side of it, then differences in osmotic pressure are quickly equalized through diffusion of the dissolved particles, and a movement of water can in consequence show itself only imperfectly.

But the cells of the alimentary mucosa are by no means equally permeable to all dissolved substances. For this reason the cells of the alimentary tract (as well as other body cells) obey the laws of osmotic pressure more perfectly when certain substances are dissolved in the fluids present in the alimentary tract than when others are concerned. A partial explanation at least of this selective permeability of cells will be given when we discuss OVERTON and MEYER's work on the lipoids.

We have yet another way of influencing the absorption of water besides through differences in osmotic pressures, and that is by *changes in the affinity of the colloids for water.* As is well known protoplasm is composed of a mixture of colloids, and experiment has shown that the amount of water absorbed by colloids can be enormously influenced by a number of external conditions. As shown by such experiments as those of HOFMEISTER,[1] gelatine plates absorb water from the chlorides of the alkali metals. They absorb much more from isotonic solutions of the bromides and nitrates of these same metals. The citrates, sulphates, and tartrates of these metals, on the other hand, cause such swollen plates to give up their water.[2] It almost seems as though in an understanding of the laws governing the absorption of liquids by colloids, and of the secretion of these same liquids when external conditions are slightly altered, lies the explanation of much that is obscure in the phenomena of absorption and secretion as they are found in the animal and vegetable organism. Changes in osmotic pressure are certainly by themselves in-

---

[1] HOFMEISTER: Archiv f. exp. Path, XXVIII, p. 210.

[2] This almost seems analogous to the secretion of water into the alimentary tract under the influence of cathartic salts.

adequate to explain more than a few of the facts observed experimentally in these fields. Perhaps pathology may also find help here.[1]

6. Lipoidal Absorption.—If a cell (such as a red blood-corpuscle) possessing a semipermeable membrane is put into a series of differently concentrated solutions of a certain substance it will show no change in volume (will not shrink or swell) in that solution which has the same osmotic pressure as the cell contents. Such a solution is said to be *isosmotic* or *isotonic* with the cell contents. If, now, solutions of different substances are prepared a certain concentration can be found for each of these in which the cell used for study

---

[1] On the hypothesis that œdema represents an increased affinity of the colloids of the tissues for water brought about through the presence in them of abnormal substances capable of increasing this affinity I have made a series of experiments which seem to support this idea. Not only does calculation show that the maximum amount of water ever held by a tissue in œdema at no time exceeds the amount easily absorbed by a gelatine plate, but acids and certain salts which increase the affinity of gelatine plates for water increase in a similar way the affinity of frog's muscles, connective tissue, etc., for water. In œdematous tissues we have not only the presence of an increased amount of CO_2, but also organic acids of various kinds, and these present in concentrations which readily increase the affinity of ordinary gelatine plates for water some 30 or 40 percent. Many poisonous substances—such as the irritant oils—which by themselves do not increase the affinity of a gelatine plate for water, nevertheless bring about an œdematous swelling when applied to living tissues, apparently because they lead to an altered metabolism of the cells concerned which brings with it the production of substances which do increase the affinity of colloids for water. An attempt to show that such substances are present in œdematous tissues lead to the experiment of introducing frog's gastrocnemii into the peritoneal cavities of normal rabbits and rabbits with an ascites due to the injection of uranium salts. In spite of the isotonicity of the normal and abnormal peritoneal fluids the frog's muscles contained in the ascitic rabbits weighed 20 to 30 percent more at the end of several days than those in the healthy rabbits. The blood-serum of an ascitic rabbit will, moreover, upon injection, bring about an œdema in a healthy one.

shows no change in volume. Since all of these solutions must then be isosmotic with the cell contents they must also be isosmotic with each other. The red blood-corpuscles (and other cells) have been employed in this way for the determination of osmotic pressure. But this is true only when the cell has a true semipermeable membrane, and such exist practically nowhere in living organisms. As the number of substances studied with reference to their power of changing the volume of red blood-corpuscles (or causing them to give up their red coloring-matter or "plasmolyzing" various plant-cells, etc.) increased it was soon found that there exist a large number of substances which affect the cells either at no concentration at all or only when present in amounts greatly exceeding the ordinary limits at which a change in volume might be expected. Attention was called to this fact when it was pointed out that the cells of the alimentary mucosa swell or shrink only when exposed to great and sudden changes in osmotic pressure through the fluids surrounding them. These substances which in solution failed to bring about changes in the volumes of cells in proportion to their osmotic pressure existed for a long time as unexplained exceptions until Overton and Meyer made a systematic study of them and so advanced most markedly our knowledge of the fundamental character of absorption and secretion.[1]

The power of a solution to abstract water from a cell (that is, to shrink or "plasmolyze" it) is dependent upon the semipermeability of the membrane surrounding the cell to the substance dissolved in the solution. If the substance is

---

[1] Overton: Vierteljahresschrift d. naturforsch. Gesellsch. in Zürich, 1895, XL, p. 1, and 1899, XLIV, p. 88; Zeitschr. f. physik. Chemie, 1897, XXII, p. 189. Meyer: Arch. f. exp. Path. u. Pharm., 1899, XLII, p. 109, and 1901, XLVI, p. 338. This account is largely taken from Höber's excellent Physikalische Chemie der Zelle und der Gewebe, Leipzig, 1902, p. 101. See also Spiro: Physikalische und physiologische Selection. Strassburg, 1897.

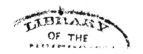

able to pass through the cell membrane plasmolysis must in consequence be impossible, for under these circumstances the diffusion of the dissolved substance into the cell equalizes the pressure on both sides of the membrane, and the difference between the osmotic pressure inside and outside of the cell which is essential for plasmolysis does not come to pass. A movement of water does not occur in the direction toward the region of higher osmotic pressure as when a semipermeable membrane is present, but a movement of dissolved particles takes place from the region of higher osmotic pressure to that of lower osmotic pressure. Only in case the cell membrane possesses but a limited permeability do both water and dissolved substance move, for in this case a cell will give up some of its water to the more concentrated solution surrounding it before the substance dissolved in this solution has had time to diffuse into the cell. Under such circumstances a *temporary* change in the volume of the cell concerned is possible. Herein lies the explanation of the behavior of the cells of the alimentary mucosa in responding only temporarily to great and sudden changes in the osmotic pressure of liquids surrounding them.

But the intensity with which a solution can bring about the plasmolysis of a cell is a function not only of the degree of difference between the osmotic pressure without and within the cell, but also of the velocity with which the substance can penetrate the cell membrane. It is evident that with the same degree of osmotic difference a substance capable of diffusing rapidly into a cell will be less likely to plasmolyze the cell than one which diffuses in more slowly, for under the latter circumstances a movement of water is more likely to occur than under the former.

It becomes possible to differentiate in consequence between substances which enter cells slowly and those which enter rapidly. Glycerine belongs to the class of substances which diffuse slowly into a cell and as slowly leave it. If algæ are in consequence placed in a dilute glycerine solution and the concentration of this is allowed to increase slowly through evaporation the algæ suffer no change, for each increase in the concentration of the glycerine solution has time to be equalized by an increase in the concentration of the glycerine within the cell. But if the algæ be removed from the now concentrated solution of glycerine and dropped into clear water they burst at once, for in so short a time the glycerine has not had time to diffuse out of the cells.

Methyl alcohol belongs, on the other hand, to the substances which can rapidly pass through cell membranes. The root hairs of *Hydrocharis* plasmolyze in a cane-sugar solution having a concentration between 7 and $7\frac{1}{2}$ percent. Plasmolysis occurs in the $7\frac{1}{2}$ percent solution in 10 seconds. If 3 percent of methyl alcohol are added to the 7 percent cane-sugar solution its osmotic pressure is made to equal a 35 percent cane-sugar solution. Yet in this mixture of sugar and methyl alcohol no plasmolysis occurs, for the methyl alcohol diffuses almost instantaneously through these cells so that the great difference between the osmotic pressure within and without the cells cannot become effective. (OVERTON.) [1]

To the compounds which diffuse rapidly into protoplasm belong the monatomic alcohols, aldehydes, and ketones, the hydrocarbons with one, two, and three chlorine atoms, the nitroalkyls, the alkylcyanides, the neutral esters of inorganic and many organic acids, anilin, etc. The diatomic alcohols and the amides of monatomic acids pass into cells more slowly, and still more slowly glycerine, urea, and erythrite. The hexatomic alcohols, the sugars with six carbon atoms (hexoses), the amino-acids, and the neutral salts of the organic acids diffuse into cells only very slowly. The entrance of these various substances into the cells is rendered apparent by yet other signs than a failure of plasmolysis, such as evidence of narcosis or other intoxication, the formation of pre-

---

[1] Cited from HÖBER: l. c., p. 105.

cipitates, etc., but the details of these experimental findings must be sought in the original publications.

A simple glance at the table given in the last paragraph shows that we have to deal with all manner of chemical substances, from those relatively simple in composition to those very complex, some of physiological importance and found as normal constituents of the living cell, others entirely foreign to the living organism. What physico-chemical character have all these substances in common which allows them to penetrate living cells more or less readily and so modify the otherwise simple osmotic behavior of cells in general?

An explanation frequently given and long believed to be the correct one is that the size of the molecules is the condition which determines the entrance of the dissolved particles. According to this conception the cell membranes may be regarded as sieves which allow all molecules that do not exceed a certain size to pass into the cell, while those larger than this are held back. The deficiencies of such an explanation are at once apparent when it is remembered that membranes which readily give passage to such large atomic aggregates as the alkaloids or sodium salicylate hold back the much simpler amino acids and potassium sulphate.

According to OVERTON *all the substances enumerated above enter cells because the membranes surrounding them behave like films composed of a substance which in its properties as a solvent is not unlike ether or the fatty oils.* For this reason all those substances which are more soluble in such ethereal or oil-like substances than in water enter the cell and this the more rapidly the greater the solubility of the substance in the ethereal or oil-like substances as compared with water; in other words, the greater the *distribution coefficient* between the film surrounding the cell and water. This distribution coefficient is, at the same temperature, independent of the concentration of the substance. This means that if a substance soluble in any two solvents is offered these simultaneously (for example, succinic acid to a mixture of equal parts of water and ether) the proportion of this substance found in solution in each of the solvents will always be the same no matter how much of the substance is offered the two solvents. This means that if of six grams of succinic acid offered a certain volume of ether and water, five grams are found to dissolve in the ether and one in the water, then if twelve grams be offered the two solvents, ten will dissolve in the ether and two in the water. The proportion of 5 : 1 is therefore maintained.[1]

With these ideas in mind it is only necessary to reëxamine the list of substances which experiment has shown enter cells more or less rapidly and see if they are not all of a character which are more soluble in ethereal or oily substances than in water, and that those which stand first in the list and consequently enter cells most rapidly are not such as have the highest distribution coefficients. An illustration may make this clearer. The repeated substitution of an atom or a group of atoms for some other atom or group of atoms in a chemical compound is often accompanied by marked changes in the solubility of this compound and its derivatives. Glycerine enters a cell only very slowly. When an atom of chlorine is introduced into this compound it enters protoplasm more rapidly, and when two are introduced still more rapidly, for these derivatives are more readily soluble in fats than the original glycerine. The same holds true of urea and its methylated derivatives. While urea diffuses but slowly into cells, the introduction of one, two, or three methyl radicles into this compound increases progressively its solubility in fats and hence the rate of diffusion into living cells.

Having shown that the entrance of many different substances into cells is dependent upon their solubility in fat-like bodies we may ask more specifically, What are these

---

[1] See HÖBER: l. c., p. 111.

substances in the cell? By making use of the so-called[?] "vital" staining methods, that is through use of stains which will color living protoplasm, OVERTON has been able to show that these fat-like bodies—or as they are collectively called, the LIPOIDS—are *cholesterin*, *lecithin*, *protagon*, and *cerebrin*. These substances are not, of course, true fats, but they resemble these in their property of dissolving more or less readily the compounds which were found above to enter living cells. The conception that living cells are surrounded by lipoidal membranes seems, therefore, not to lack experimental support, and osmotic pressure as a force determining the movement of dissolved particles and of water becomes a more clearly defined force in phenomena of absorption and secretion when this selective power of solution of the surface films of cells is taken into consideration.

It must not be thought, of course, that these conceptions of cell membranes and the movement of water and dissolved substances through them, as outlined for cells in general and applicable to the alimentary mucosa in particular, explain more than a part of the phenomena of absorption and secretion as illustrated by the gastro-intestinal tract. Even in this statement we are not considering the absorption or secretion of substances which before or during their passage through the alimentary mucosa suffer a change, such as the fats, proteins, and carbohydrates. All this will become more apparent in the discussion of the absorption of the specific elements of our food, when a mass of isolated facts will be found that still lack a unifying explanation.

7. The Absorption of Salts.—Recognized physical laws suffice at present to explain only a part of the phenomena observed in the absorption from the gastro-intestinal tract of even such simple substances as the salts. From some points of view the absorptive mucous membrane behaves like a dead diffusion membrane on one side of which there is found the blood, the composition of which may be looked upon as fairly constant, on the other the salt solution undergoing absorption.

A salt solution will, in the course of time, be absorbed by the alimentary tract no matter whether it be isotonic, hypertonic, or hypotonic with the blood, but the rate differs with the different kinds of salts. Different observers agree that when a *hypertonic* solution is introduced into the intestine its volume increases at first through a diffusion of water into it, while its concentration diminishes. At the same time certain constituents of the blood diffuse into it. After this the volume of solution in the intestine gradually diminishes. These phenomena are usually explained on the basis that the salt diffuses into the blood because its concentration in the lumen of the alimentary tract is higher than the concentration of this same salt in the blood, while various constituents of the blood diffuse out into the intestine for the same reason. The initial increase in the volume of the solution is explained in those cases in which it occurs on the basis that the salt under consideration does not diffuse rapidly enough not to allow osmotic differences to make themselves felt through a migration of water. For this reason, too, it is found that the originally hypertonic solution gradually approaches isotonicity with the blood.

When we deal with the diffusion of a *hypotonic* solution water diffuses into the blood because of osmotic differences, and the salt solution under these circumstances also approaches isotonicity with the blood. When this occurs the concentration of the salt in the solution undergoing absorption is increased and so is placed in a position to diffuse into the blood, but this lowers the osmotic pressure of the solution in the lumen of the intestine once more, in consequence of which water is again absorbed from it by the blood; and so these processes repeat themselves until all the salt solution is absorbed.

Solutions *isotonic* with the blood are absorbed quite as easily as hypertonic or hypotonic ones. It is somewhat difficult to see what forces are active in bringing about the transport of the solution into the blood. The explanation ordinarily given is as follows: While the inorganic constituents of the blood may diffuse out into the lumen of the intestine, the organic constituents (the albumin, globulin, etc.), because of their colloidal nature, are practically unable to do this. While colloids exert no great osmotic pressure they nevertheless exert some, and so, even after all other osmotic differences on both sides of the diffusing membrane have been equalized, an excess of osmotic pressure must remain on the side of the blood. This would then lead to the abstraction of a small amount of water from the solution in the intestine which would in consequence be rendered hypertonic. Salts would then diffuse into the blood, then more water, until, little by little, the whole would be absorbed.[1]

For a large number of salts the *rate of absorption* is proportional to the velocity of their diffusion (HÖBER [2]). In the following table are arranged a number of salts in the order of their diffusion velocities. When arranged in the order of the velocity with which these salts are absorbed from isotonic solutions when equal amounts are introduced into closed loops of intestine the grouping remains the same.

Sodium chloride

Sodium nitrate
Sodium lactate

Sodium sulphate
Sodium malonate
Sodium succinate
Sodium tartrate
Sodium malate

Magnesium chloride
Calcium chloride

[1] More detailed discussion of these still unsatisfactory theories of absorption cannot be entered into here. See HÖBER: Physikalische Chemie d. Zelle u. d. Gewebe, Leipzig, 1902, p. 184; PFLÜGER'S Archiv, 1898, LXX, p. 624. STARLING: Journal of Physiology, 1896, XIX, p. 313. KÖVESI: Centralbl. f. Physiol., 1897, XI, p. 553.

[2] HÖBER: PFLÜGER'S Archiv, 1899, LXXIV, p. 246. Zelle und Gewebe, Leipzig, 1902, p. 190.

Sodium chloride has, of all these salts, the greatest diffusion velocity, and also the greatest absorption velocity. From here downwards the velocities of diffusion and of absorption diminish progressively. In the same unit of time a larger amount of the solution of a salt in the first group will disappear from a loop of intestine than of an isotonic solution of a salt contained in the second group, and this will disappear sooner than an equal amount of an isotonic solution of a salt found in the third or fourth group. For all salts this does not hold, however. Of the halogen salts of sodium, for example, which have all the same diffusion velocity, the chloride is absorbed most rapidly, then the bromide, and finally the iodide. All fluorides are absorbed exceedingly slowly, as also the carbonates, due no doubt to secondary changes produced in the cells, for the fluorides are protoplasmic poisons, and the carbonates suffer a hydrolytic dissociation with the formation of free OH ions, which are toxic in even very low concentrations.

HÖBER has made interesting observations on the *paths of absorption* of the salts. From studies with dyes which are in part soluble in the lipoids of the cells, in part insoluble, he has been able to show that salts are for the most part absorbed only intercellularly; that is to say, they pass into the blood not through the epithelial cells of the absorbing mucosa, but through the intercellular spaces. If solutions of the salts of various dyes soluble in the lipoids of the cell—such as methylene blue, toluidin blue, or neutral red—are introduced into the intestinal tract of frogs, microscopic examination shows that the dyes affect slightly all portions of the absorbing mucosa. But while protoplasm, nucleus, and intercellular substance are scarcely colored, granules contained within the protoplasm take up the dyes most intensely. The granular material seems to be, therefore, an excellent solvent for these dyes and probably consists of those substances which are collected under the heading lipoids. When solutions of dyes insoluble in the lipoids—such as water-soluble aniline blue, water-soluble nigrosin, or benzoazurin—are introduced into the intestine the urine becomes colored, yet no deeply stained granules are found in the protoplasm.

These observations only show that the intestinal epithelium is permeable to salts soluble in the lipoids; it does not as yet prove that those insoluble in the lipoids are absorbed only interepithelially. But this is rendered probable as soon as fixing agents, such as ammonium molybdate, osmic acid, corrosive sublimate, picric acid, ammonium picrate, tannic acid, or gold chloride, are used to fix the pictures obtained after simple introduction of a dye into the intestinal tract. Under these circumstances it is found that if the fixing agent is one soluble in the lipoids of the cell (such as osmic acid), the granular pigmentation of the cell found after the use of such a dye as methylene blue remains unchanged. If, however, a fixing agent insoluble in the cell lipoids (such as ammonium molybdate, which is able to plasmolyze cells) is used, the blue granules in the cell are seen to dissolve, to move toward the periphery of the cell and to be precipitated here. This is because the ammonium molybdate remains in the intercellular spaces and precipitates the dye present here. In this way the equilibrium between the dye without and within the cells is destroyed. The stain in consequence begins to move out of the cell toward its periphery, where it meets the ammonium molybdate. What has been said of this fixing agent holds for every one of the fixing agents capable of reacting with the stains employed and not soluble in the lipoids. With this is proven quite conclusively that salts insoluble in the lipoids make their way from the intestine into the blood only through the interepithelial spaces.[1]

This is, perhaps, the best place to touch briefly upon the absorption of such compounds as alcohol, urea, and certain

---

[1] It might seem from this that the ordinary salts are entirely incapable of entering the epithelial cells of the intestine or any other cell containing lipoids. This is not true, but the means by which they may or do enter cannot be dealt with here.

other substances which are absorbed from the alimentary tract much more rapidly than the inorganic salts. While the absorption of salts from the stomach is still questioned, the rapidity with which alcohol disappears from the gastric contents is truly phenomenal. Urea also is much more rapidly absorbed than salts of a simpler composition. The reason for this is at once apparent when we assume that while the inorganic salts can pass into the blood only through the intercellular spaces, alcohol, urea, etc., are soluble in the lipoids, and so can pass directly through the epithelial cells as well.

From what has been said it seems, therefore, as though the behavior of the epithelial cells of the gastro-intestinal tract toward certain soluble substances is not unlike the behavior of cells in general, and is dependent upon the same powers of selective solution by their surface membranes.

We come now, however, to a series of facts for which an adequate physical explanation is entirely lacking. These facts are connected with the *predominant permeability of the gastro-intestinal tract in one direction.* While, as already stated, some of the constituents of the blood may diffuse into a solution contained in the lumen of the intestine, only very little passes out in this way. Salts found in the blood pass only in exceedingly small amounts, if at all, into the intestine, while these same salts in various concentrations pass easily from the alimentary lumen into the blood. It is evident that every constituent of the blood, which, like the albumin and the globulin, is unable to pass into the intestine can in this way become effective in absorbing water, and in consequence lead to a passage of salt from the fluid in the alimentary tract into the blood. This semi-permeability of the absorbing mucosa in one direction only would therefore greatly favor the absorption of substances from the lumen (COHNHEIM).

All our attempted physical explanations of absorption by the alimentary mucosa fail when we approach the observations of HEIDENHAIN,[1] REID,[2] and COHNHEIM. HEIDENHAIN found that when a dog's own blood-serum is introduced into its intestinal tract it is absorbed. REID removed the intestine from rabbits at the height of digestion and found when this is cut open and stretched between two isotonic sodium chloride solutions that the salt solution is pumped from the side of the mucosa toward the serosa. COHNHEIM found that when the alimentary tract is removed from certain marine animals (*Holothuria tubulosa*), and this is filled with 10 to 30 c.c. of sea-water, and the whole is then suspended in sea-water, that this moves from the alimentary tract out into the surrounding water. In all these experiments we have no differences in osmotic pressure, and at present we cannot say what makes the liquids move. The death of the cells, or certain poisons such as sodium fluoride, arsenic, or chloroform, do away with this transport of serum, sodium chloride solution, or sea-water from the side of the mucous membrane toward the serosa, and make the alimentary wall act like an ordinary dead diffusion membrane. But it explains nothing, of course, when we say that such a transport is dependent upon the living activity of the cells.

The predominant permeability of the intestinal tract in one direction can be markedly influenced experimentally. When a dextrose solution is introduced into a loop of small intestine, it is absorbed. According to experiments carried out by GERTRUDE MOORE and myself this same glucose solution when injected into the blood is not excreted into the lumen of the intestine. This happens, however, as soon as certain salts (such as sodium chloride) are injected along with the sugar solution, presumably because these salts modify this predominant permeability in one direction. The intestine in consequence now excretes a substance which it

---

[1] HEIDENHAIN: PFLÜGER'S Archiv, 1894, LVI, p. 579.

[2] REID: Phil. Trans. Royal Soc., 1900, CXCII, p. 231. Journal of Physiol., 1901, XXVI, p. 436.

absorbed formerly. What has been said of dextrose holds also for a number of other substances.

The different portions of the alimentary tract absorb salts in very different amounts. Under ordinary circumstances the amount of salt absorbed in the mouth or œsophagus is to be looked upon as practically nothing. While certain authors believe that considerable amounts of salt are absorbed in the stomach, others question it entirely. The small intestine absorbs salts freely throughout its entire length, though the jejunun seems to be somewhat more active in this regard than the ileum. The large intestine also takes up salts, but not to the same extent as the small bowel.

Under certain circumstances one region of the alimentary tract may absorb a salt while another is excreting this same salt. One and the same portion of the intestine may even absorb a salt which under somewhat different conditions it excretes. The character of the changes which take place in the absorbing structure of the alimentary tract to render such phenomena possible are not as yet understood, and the visible alterations (swelling, contraction, coagulation) often observed in the absorbing surfaces still lack a unifying physical explanation.

CHAPTER XVI.

THE ALIMENTARY TRACT AS AN ABSORPTIVE SYSTEM
(Continued).

8. The Absorption of Carbohydrates.—The carbohydrates which are of chief importance in the physiology of alimentation in so far as they are found in any ordinary mixed meal, are the polysaccharides starch, glycogen, and cellulose, the disaccharides cane-sugar, malt-sugar, and milk-sugar, and the monosaccharides dextrose, lævulose, and galactose. Of the polysaccharides the starches make up not only the bulk of this class of food, but of all the carbohydrates that are enjoyed in an ordinary diet. The consumption by an ordinary individual of several hundred grams of starch a day in the form of bread, potatoes, beans, etc., is not uncommon. A few grams of glycogen enter into the ordinary daily diet as constituents of lean meat, liver, etc. Cellulose is obtained through the vegetable constituents of the diet, more particularly celery, string beans, turnips, carrots, beets, etc.

Sucrose (cane- or beet-sugar) as the ordinary sugar of commerce and the recognized sweetening agent of our food makes up the bulk of the disaccharides which we consume, though the exact amount consumed in a day is subject to the widest individual variations. Malt-sugar is obtained in small amounts through beverages and "breakfast foods," into the composition of which sprouted grains enter. Milk-sugar comes to us through milk and certain of its derivatives; only rarely as a distinct addition to an ordinary mixed diet.

Dextrose and lævulose are found in certain fresh and dried

282

fruits, and in the commercial "glucoses" (corn-sugar, molas-ses, etc.). The dextrose and lævulose found in the alimentary tract is derived, with the exceptions indicated, from the de-composition of the polysaccharides and disaccharides of the diet. Galactose enters as such into the diet practically not at all. It is produced in the alimentary tract through the action of lactase upon milk-sugar. Either directly or indirectly dex-trose becomes the predominating sugar of the alimentary tract. Not only does it constitute the bulk of the com-mercial "glucose," but it appears as the ultimate product of the decomposition of every polysaccharide and disaccharide. Glycogen is split directly into dextrose through amylase. The same ferment converts starch into maltose, which through maltase is converted into dextrose. Dextrose appears as one half of the product of the action of sucrase on sucrose, and lactase on lactose. Cellulose, though acted upon only by the bacterial enzymes of the alimentary tract, yields dextrose when this occurs.

We have to ask now in what form the carbohydrates of the diet are absorbed. There seems to be little doubt that the monosaccharides are absorbed as such. With the disac-charides matters are somewhat different. If sucrose, maltose, or lactose are fed slowly, they are all converted into mono-saccharides before they are absorbed. When, however, these disaccharides are fed rapidly, then they are not all split before they are absorbed, and sucrose, maltose, and lactose may be recovered as such from the blood and from the urine, for these disaccharides when present in the circulating blood in a concentration exceeding a very small fraction of a per-cent are eliminated through the kidneys. Ordinarily it is said that the disaccharides appear in the blood and the urine if they are fed in too large *amounts*. This is not the essential factor, however, but the *time* taken in feeding the amount, for if the sugar solution is not absorbed too rapidly it does not pass over into the blood in an unchanged state.

Starch, glycogen, and cellulose all being colloidal bodies are unable as such to pass through the walls of the alimentary tract. They must be acted upon by the enzymes found here before they can be absorbed. The starch may be absorbed as soon as it has been broken down to the maltose stage, but most of it seems to be absorbed in the form of dextrose. The glycogen is readily converted into dextrose and is as rapidly absorbed. The cellulose of the food is ordinarily classed as a substance which is of no use from a nutritional standpoint, for it cannot be absorbed as such and no enzymes are secreted by the alimentary tract which can act upon it. Only the cellulose-splitting ferment (cytase) contained in certain of the bacteria found in the alimentary tract is able to convert cellulose into sugar, and experiment shows that the amount produced in this way is exceedingly small.

Of great interest from a medical standpoint are the experiments of HOFMEISTER,[1] who has determined the "assimilation limits" of various carbohydrates. By the assimilation limit (which is not a well-chosen term) HOFMEISTER understands the amount of a sugar that may be administered to an animal without the appearance of sugar in any form in the urine. It is clear that a large number of factors play a rôle in this complicated picture, and it is not strange that the figures obtained should vary widely from each other. Nevertheless the general conclusions which may be drawn are clear enough and are valuable in any dietary scheme in which the quality and quantity of carbohydrates administered plays a rôle. It has been found that the assimilation limit of any sugar is different not only in different animals, but varies in one and the same animal under different physiological conditions, such as the rapidity with which absorption takes place, the amount of sugar already present in the blood, and the nutritional state of the animal as a whole. A dog that has been starved for a number of days will excrete sugar in the

---

[1] HOFMEISTER: Arch. f. exp. Path. u. Pharm., 1889, XXV, p. 240, and 1890, XXVI, p. 350.

▶ urine after being fed 10 to 15 grams of starch paste, while a healthy dog will stand several times this amount at a single feeding and not excrete sugar.[1] Most striking is the assimi- · lation limit of one and the same animal for different kinds of carbohydrates. Ordinary starch leads all the others in the amount that may be consumed in twenty-four hours (more than 500 grams) without the appearance of sugar in the urine. Next in order stand dextrose, lævulose, sucrose, maltose, and lactose in the order named. The reason why the disaccharides stand lowest is no doubt to be sought in the fact that when these are fed rapidly and in large amounts, they pass without change into the blood, and since the liver and muscles retain the disaccharides but imperfectly their concentration in the blood soon exceeds the limits at which the sugar passes over into the urine.

The rapidity with which the different sugars are absorbed seems not at all dependent upon their rates of diffusion or differences in osmotic pressure. The subject has been investigated by ALBERTONI,[2] RÖHMANN and NAGANO.[3] Sugars are absorbed not only from solutions which are hypertonic or isotonic with the blood but also such as are hypotonic. When isosmotic solutions of sugars are compared it is found that dextrose is absorbed most rapidly, sucrose next, and much more slowly lactose. But the amount absorbed in any unit of time varies, much more of any given sugar being absorbed in the first hour after feeding than in subsequent hours. Starch, as already pointed out, cannot be absorbed as such. Its absorption is dependent upon the velocity with which it is split into absorbable sugars. These sugars under normal circumstances are absorbed as rapidly as formed. Since the amount of amylolytic change is greatest not immediately after eating but in the second or third hour of the digestive

---

[1] My own attempts to repeat this experiment succeeded only once on one out of four dogs.

[2] ALBERTONI: Centralbl. f. Physiol., 1901, XV, p. 457.

[3] RÖHMANN and NAGANO: Centralbl. f. Physiol., 1901, XV, p. 494.

period, it is found that starch is also most rapidly absorbed during this period.

After what has been said it becomes somewhat difficult to explain the fate of starches which under experimental conditions may be fed an animal in which the salivary and pancreatic secretions are lacking, or patients in whom these are insufficient in amount or poor in the proper enzymes. It has been found that in dogs in which the pancreatic ducts have been ligated that upwards of 50 percent of the starches fed cannot be recovered from the fæces. By what agencies these starches are rendered absorbable is not known, for the action of the gastric acid, the bacteria of the alimentary tract, and the slight amylolytic activity attributed by some to the intestinal juice seem insufficient.[1]

The question of the channels through which the carbohydrates are distributed to the body after passing through the epithelium of the alimentary tract seems to be settled beyond question. v. MERING showed in 1877 that the absorbed carbohydrates pass through the portal vein to the liver, for while under normal conditions the blood of this vessel contains no more than about 0.2 percent dextrose, it may contain twice this amount after a carbohydrate meal. Determinations of the sugar content of the lymph obtained from the thoracic duct showed that this fluid contained, both before and after a meal of carbohydrates, a fairly constant percent (less than 0.2 percent) of sugar. This shows that under ordinary circumstances all the carbohydrates of the food leave the alimentary tract by way of the blood. When, however, excessive amounts of carbohydrates are fed a small portion of them may pass over into the lymph-channels, as shown by GINSBERG's observations on rabbits and dogs. MUNK and ROSENSTEIN's studies on a case of lymphatic fistula in a human being fully confirm these observations. It was found in this case, in which practically all the lymph

---

[1] See MUNK: Ergebnisse d. Physiol., 1902, I, 1te Abth., p. 308.

coming from the intestines was secreted externally, that less than $1/2$ gram of sugar was excreted through this channel after a feeding of 100 grams of starch and sugar, in other words, not even 1 percent of the absorbed carbohydrates.[1]

9. The Absorption of Fat.—The means by which the often enormous quantities of fat—100 to 200 grams daily— which the human being consumes are absorbed has for more ⸱ than half a century been the subject of most active research and discussion.   It is clear that in the fat of our food we have a substance which is, practically speaking, insoluble in water and but little more soluble in protoplasm. In discussing the problem of fat absorption we must, therefore, ask first of all, Can fat be absorbed as such or is it first changed into a soluble form?

Those authors who have held to the idea that fat is absorbed as fat have called attention to the fact that even though it is insoluble, fat can be very finely divided and kept suspended in water in a so-called emulsion. Not only do many fats enter the alimentary tract in the form of such an emulsion—for example, milk and raw yolk of eggs—but a number of agents exist in the body which have the power of emulsifying fats to a very high degree. As of first importance in this direction we must mention the bile, under the influence of which any of the ordinary fats of the diet, such as butter, the fat of meat, and olive-oil, may be speedily emulsified.·

While these agents exist which can alter so markedly the *physical* state of the fats, there are others which are equally potent in bringing about a *chemical* change in them. Of first importance in this direction is lipase, a widely distributed ferment which has the power of splitting fats into fatty acid and alcohol. This chemical change which fats may suffer in the body affects markedly the question of fat absorption, for these products of fat digestion are many of them soluble

---

[1] See MUNK: Ergebnisse d. Physiol., 1902, I, 1te Abth., p. 309, where references to the literature will be found.

in water. Others are soluble in water if bile is present, and some are soluble in protoplasm. If, therefore, it could be shown that under the conditions which exist in the body it is possible for all the fat of a fatty meal to be converted into digestion products soluble in the body fluids, and this within the time allowed for the absorption of such a meal under physiological conditions, one of the great difficulties in the way of believing that all fat is absorbed only in the form of its soluble digestion products would be overcome. The extent to which the fat of a fatty meal is split into fatty acid and alcohol during an ordinary period of digestion has been greatly underrated by the majority of investigators. We know now from the observations of RACHFORD [1] that in the hours making up the ordinary period of pancreatic activity consequent upon a meal ample time is allowed for the splitting of at least the larger portion, and probably all of the fat consumed in that meal.

Before entering further into the discussion of this question let us ask first of all whether fat *can* be absorbed in the form of an emulsion. If this question is answered in the affirmative, then we have to ask, Is all the fat absorbed in this form, or only a part of it, and how much? The mere question as to whether fat *can* be absorbed in the form of an emulsion must, perhaps, be answered in the affirmative. From a physical standpoint the question is one which asks whether substances having a high molecular weight can diffuse through animal membranes. The physical experiment to settle this question has been made by EIJKMANN,[2] who found that a solution of glue poured upon an agar-agar plate will, if a proper temperature be maintained, soon diffuse into the agar-agar. In this case we have one colloid diffusing into another. The physiological experiment to answer this question has been made by FRIEDENTHAL,[3] who fed colloidal so-

---

[1] RACHFORD: Journal of Physiology, 1891, XII, p. 72.

[2] EIJKMANN: Centralbl. f. Bacteriol., 1901, XXIX, p. 841.

[3] FRIEDENTHAL: Archiv für (Anat. und) Physiologie, 1902, p. 149.

dium silicate to rabbits and young dogs and was able to re_ cover the salt from the urine in exceedingly small but never_ theless distinct amounts. This means, of course, that the colloidal sodium silicate diffused through the gastro-intestinal wall.

There is a difference, however, between asking whether fat can pass as such through the intestinal mucous membrane and whether under ordinary circumstances this *is* the way in which it is absorbed. ' There seems to be little doubt that even if fat can be absorbed as such, most of it passes through the intestinal wall (and from tissue to tissue) in the form of its soluble digestion products.' As evidence of this we may quote the experiments of CONNSTEIN.[1] If the essential change necessary for the absorption of fats lay in their conversion into an emulsion in the animal body, then it would be reason- able to expect that a fine emulsion of one fat should be absorbed as rapidly as an emulsion of any other fat provided it were equally finely divided. An emulsion of lanolin ought, therefore, to be absorbed as readily as an emulsion of butter- fat. As an actual matter of fact, CONNSTEIN found that when he fed a dog with a lanolin emulsion in water 97½ percent of the entire amount fed could be recovered from the fæces. ' The formation of an emulsion from the fat is, therefore, only of secondary importance in the absorption of this foodstuff. ' The real reason why lanolin is not absorbed in the above ex- periment lies in the fact that it is acted upon only exceedingly slowly by the fat-splitting enzymes of the digestive tract, and hence is not converted into the absorbable products of fat digestion.

The recognition by KASTLE and LOEVENHART, and inde- pendently of them by HANRIOT, that the action of lipase is re- versible, has altered entirely our conception of the mechanism by which fat is absorbed from the intestinal tract.[2] The belief

---

[1] CONNSTEIN: Archiv für (Anat. u.) Physiol., 1899, p. 30.

[2] See p. 151, and LOEVENHART: American Journal of Physiology, 1902, VI, p. 332.

of the older observers that fat is absorbed chiefly in the form
of an emulsion was based upon physiological experiments
and histological studies which showed that in ordinary digestion the fat of the food is emulsified in the lumen of the gut,
appears as droplets in the cells lining the gut, and in the same
form in the chyle, the fat-laden lymph which leaves the intestinal tract after a fatty meal. This conception was further
strengthened by the analytical results of physiological chemists, who found but little fatty acid and alcohol (glycerine) in
any of these localities. The idea, therefore, that fat was absorbed in any other form than fat received little support, and
the formation of fatty acid and alcohol during digestion was
looked upon as of little importance.

As the following will show, fat is in all probability absorbed
only after it has first been split into fatty acid and alcohol
and never in the form of the original foodstuff. The observations of RACHFORD and the recognition of the reversible
action of lipase suffice entirely to explain the facts cited
above which have so long been quoted in support of the older
ideas of fat absorption.

Under normal circumstances fat is absorbed with great
rapidity from the intestinal lumen. From a physico-chemical
standpoint alone, therefore, it appears very unlikely that fat
enters the cells of the intestinal mucosa as such, for fat has
only slight powers of diffusion, especially into solvents made
up chiefly of water, such as protoplasm. To overcome this
argument the intestinal epithelium has been endowed with
powers of amœboid motion and (hypothetical) tubules surrounded by contractile protoplasm. Even were its entrance
into the intestinal mucosa explained in this way, its exit into
the lacteals and from here into the various tissues of the body
would yet have to be accounted for. The same explanation
could evidently not hold in all these cases of absorption. The
belief, on the other hand, that fat enters and leaves the intestinal mucosa only in the form of its cleavage products—in fact,
enters and leaves all tissues in this form—meets with no such

objections. The cleavage products of fat are readily soluble in the fluids and tissues of the body and diffuse with great rapidity.

The exact form in which the cleavage products are absorbed cannot as yet be looked upon as settled definitely. For the sake of simplicity we will adopt MUNK and LOEVENHART's opinion that fat is absorbed as fatty acid and alcohol. This is probably correct, though it is at variance with the belief of certain other investigators that the fatty acid unites with the alkalies of the body and forms soaps. Whichever may ultimately have to be adopted will not alter the fundamental principles of the mechanism of fat absorption. Since certain of the fatty acids are practically insoluble in water, and since solubility is so important a factor in absorption, it is well to bear in mind MOORE and ROCKWOOD's experiments, which show that the insoluble fatty acids become freely soluble in the presence of bile.

It is well to impress anew upon the reader that lipase, which we ordinarily think of as a constituent of the pancreas and the pancreatic juice only, is really very widely distributed throughout the body, where it occurs in different amounts in practically every tissue and fluid. Of immediate interest to us is the fact that lipase occurs in the mucosa of the intestine as also in the lymphatic glands, lymph, and the blood.

Bearing in mind the facts stated above regarding lipase, its action and its distribution, let us trace the chemical changes which follow the ingestion of a fat. This substance is not acted upon in the mouth or œsophagus. As lipase is found in the gastric juice and gastric mucosa it is possible that some digestion may occur in this viscus. This will occur especially if the meal has been one which calls forth a secretion of but little hydrochloric acid or if the digestion occurs in a stomach which through pathological change is secreting a deficient amount of this acid. The activity of the lipase under these circumstances is not interfered with. It is entirely possible that the appearance of butyric and other fatty acids in diseases of the stomach is attributable in part at least to the activity of the lipase normally present here.

But even if under normal circumstances little or no digestion of fats occurs in the stomach, active digestion of this food begins as soon as the gastric contents are neutralized in the duodenum and have poured out upon them the pancreatic juice and the bile. The function of the latter we will ignore for the present. Let us ask first of all how much of the fat will be split under the influence of the pancreatic juice as the mixture of the two moves down the intestine. It is clear that if the intestine were replaced by a glass tube, by no means all of the fat would be split, but only a portion of it, or to put it more technically, fat would undergo cleavage until an equilibrium had been established between this substance on the one hand and fatty acid and alcohol on the other. In other words, we recognize here again an equation similar to the one given on page 107:

$$\text{Fat} \rightleftarrows \text{Fatty acid} + \text{Alcohol.}$$

In the animal body conditions are somewhat different than in a glass tube. While in a glass tube the products of the cleavage accumulate, this does not occur in the intestine, for here the fatty acid and alcohol diffuse into the intestinal mucosa as soon as formed. Evidently, therefore, the state of equilibrium outlined above never comes to pass in the intestine, and the cleavage of the fat continues until all has been split, in other words, all has been absorbed. This really occurs, for under normal circumstances no fat, or practically none, is found in the fæces.

Let us see now what becomes of the fatty acid and alcohol which have diffused into the lining cells of the intestine. At the beginning of a meal these cells contain no fat, but they do contain lipase. As the fatty acid and alcohol diffuse into them evidently the reverse of what occurred in the lumen of the gut must happen here, for since the action of lipase is reversible it must synthesize fat from the products of the fat digestion which is going on in the lumen of the intestine. In other words,

$$\text{Fatty acid} + \text{Alcohol} \rightleftharpoons \text{Fat.}$$

It is this synthesis of fat in the epithelium of the gut which gives rise to the appearance of fat droplets in this locality, and which the older observers looked upon as evidence supporting the idea that fat is absorbed as an emulsion.

We have yet to explain the transport of the fat into the lymph-channels. The fatty acid and alcohol which diffuse into the cells lining the intestine do not stop here but pass on into the lymph current beyond. At the beginning of a digestion period this is also free from fat, but it contains lipase. Evidently the same play must occur in this liquid tissue which we saw take place in the cells lining the intestine. Under the influence of the lipase the fatty acid and alcohol are synthesized into fat, and it is this fat which evidences itself in the droplets found in the lymph returning from the intestine. That we are dealing with a chemical equilibrium in this case also is shown by the fact that the lymph always shows the presence of free fatty acid accompanying the fat.

In order to get a conception of the process of fat absorption as a whole, we need only to imagine these different processes of analysis, diffusion, and synthesis going on simultaneously and side by side. We are in a position now to recognize the unimportant rôle played by the fat droplets which appear in the epithelial cells of the intestine. They are evidently transitory creations, which exist only as long as fatty acid and alcohol are diffusing through these cells. When the last remnants of fatty acid and alcohol diffuse out of the cells into the lymph stream the fat in the cells can no longer maintain itself, and the lipase which before hastened its synthesis now hastens its analysis until it too has disappeared as fatty acid and alcohol into the lymph stream. This leaves the cells once more in the empty condition in which they were found at the beginning of the digestion period.

Under the influence of the lipase they contain, the tissues of the body also store fat when it is plentiful in the food and give it up during starvation in the same manner as the epithelial cells of the intestinal mucosa; but a discussion of this problem is beyond the limits of our subject. One more fact is of interest in support of the ideas of fat absorption advanced above. We would expect from chemical considerations alone that if a certain fat is digested by lipase, the products of its digestion should, when synthesized under the influence of the same ferment, yield the original fat, and that this ought to hold when different kinds of fat are consumed by animals. Not only were MUNK and ROSENSTEIN able to isolate from the lymph obtained from their patient with a lymphatic fistula an oily fat when olive-oil was fed, and a tallow-like fat when mutton-tallow was administered, but ROSENFELD[1] has been able to show more recently that if one dog is fed cocoa-butter and another mutton-tallow the fats deposited in each case correspond to those ingested. Goldfish and carp, moreover, deposit mutton-tallow when this is fed to them.

Unlike the absorption of the carbohydrates and the proteins, which leave the alimentary tract with the blood stream, the fats leave the absorptive mucous membrane almost entirely by way of the lymph. Very shortly after a fatty meal the lymph leaving the intestine assumes a milky appearance, owing to the exceedingly fine droplets of fat contained in it, and at the height of absorption may contain from 3 to 8 percent of fat. The lymph returning from the intestine and thus laden with fat is known as *chyle*. MUNK and ROSENSTEIN found that more than 60 percent of the fat consumed by their patient with a lymphatic fistula could be recovered from the lymph within the first twelve hours after feeding.

---

[1] ROSENFELD: Verhandlungen d. XVII. Congress f. innere Medicin, 1899, p. 503.

Some fat seems, however, to be carried from the intestine by way of the blood-vessels, and when the thoracic duct has been ligated the amount carried in this way is much increased. MUNK and FRIEDENTHAL found that the fat in the blood may under circumstances rise to six times the normal amount when the thoracic duct is occluded. A thick layer of fat forms upon the blood collected from a vein even though the entire amount of fat absorbed from the intestine may be less than half that absorbed when the thoracic duct is open.

The *total amount* of fat absorbed is dependent upon the kind of fat fed an animal. We know from MUNK and MÜLLER's observations that fats having a low melting-point are more perfectly absorbed than those having a higher one. 97.7 percent of the olive-oil fed an animal is absorbed. 97.5 percent of fats melting between 25° and 34° C. (goose- and pork-fat), 90 to 92.5 percent of fats melting between 49° and 51° C. (mutton-tallow), and only 89.4 percent of a mixture of stearin and almond-oil melting at 55° C. are absorbed. Of pure stearin, which does not melt until 60° C. is reached, only 9 to 14 percent is absorbed.[1] When a mixture of fats having different melting-points is fed, those having the lower melting-points are absorbed more completely than those having the higher. Fat is also more perfectly absorbed when it is free in the form of butter or lard than when it is enclosed in cells, as in fat meat or bacon. Under these circumstances the digestive juices must first dissolve the walls of the cells.

The velocity with which different fats are absorbed is also determined in large part by their melting-points. MUNK and ROSENSTEIN found that the lymph obtained from their patient with a lymphatic fistula became milky in appearance two hours after feeding a mixture of olive-oil containing 6 percent oleic acid, and in the fifth hour after feeding con-

---

[1] MUNK: Ergebnisse d. Physiol., 1902, I, 1te Abth., p. 323.

tained a maximum of 4⅓ percent fat. When mutton-tallow was fed a maximum of 3.8 percent fat in the lymph was not reached until the seventh or eighth hour. The consumption of a fat melting at 53° C. did not lead to more than a milkiness of the lymph in the fifth to the sixth hour (only 0.7 percent fat), which continued with a progressive fall in the percentage of fat until the thirteenth hour.

This is, perhaps, the best place at which to discuss the importance of the bile and the pancreas in the absorption of fats. This question is of clinical moment, for a deficient secretion or total lack of bile or pancreatic juice is encountered in a number of pathological conditions. All authors are agreed that when experimentally or through pathological states the bile is prevented from entering the duodenum the fats are much less perfectly absorbed than under normal circumstances. This fact was recognized half a century ago by BIDDER and SCHMIDT, and has since then been corroborated and amplified by a score of investigators. The exact amount of fat absorbed by men or animals when no bile enters the intestine is subject to great variation, so that it is not surprising that the figures of different students of the question vary greatly. It is ordinarily stated that less than half of the fat which under normal circumstances readily disappears from the alimentary tract is absorbed when no bile is present. The highest figures ever attained are the disappearance of 70 percent of the fat against 95 percent in normal animals as determined by analysis of the fæces. The character of the consumed fat plays an important rôle. A fat with a low melting-point is more readily absorbed than one with a higher one, just as under normal circumstances.[1] MUNK found that the absorption of mutton-tallow, for example, was interfered with twice as much as the absorption of pork-fat. A further interesting fact which at present still lacks an explanation is that the proportion

---

[1] MUNK: Ergebnisse d. Physiol., 1902, I, 1te Abth., p. 324.

of fatty acids to fat in the fæces is much higher when no bile is present than when it is. Normally, two-thirds to seven-tenths of the fatty substances present in the fæces are fatty acids, while in the absence of bile eight-tenths to nine-tenths are fatty acids, the rest neutral fat. It was formerly believed that this could be explained through MOORE and ROCK-WOOD's observation that the fatty acids are more soluble in the presence of bile and hence are more readily absorbed when this secretion is present, but MUNK has shown that the reverse is true by finding that fatty acids are more readily absorbed than neutral fats in the absence of bile.

The many observations at hand on the effect of removal of the pancreas or ligation of its duct all agree that the absorption of fat is much hampered by such procedures. Vastly different figures are, however, found in the literature to illustrate the extent to which the absorption of fats is decreased. When not in the form of an emulsion fats are absorbed in very slight measure when the pancreatic secretion is entirely lacking. HÉDON and VILLE state that 10 percent of an olive-oil feeding may be absorbed and 22 percent of the fat of milk. MINKOWSKI and ABELMANN give even higher figures for the absorption of oil in the form of an emulsion, 28 to 53 percent of milk-fat.[1] That any fat at all should be absorbed in the absence of a pancreatic secretion was long regarded as a remarkable fact. This can no longer be considered a mystery since we have become acquainted with the almost universal distribution of lipase in the tissues and fluids of the body. The mucous membrane of the small intestine contains this ferment, and no doubt some is found in the secretions of this viscus. To the ferment present from these sources must be attributed whatever digestion and absorption of fat we may encounter in an animal deprived of its pancreatic secretion. With the limited amount of lipase present under these circumstances it cannot seem

---

[1] Quoted from MUNK, l. c., p. 325.

strange that an emulsion of a fat with its larger surface should be better absorbed than the same fat when not previously emulsified. It is self-evident that with a deficient amount of lipase the conditions normally provided for the rapid production of an emulsion from the fat of the food are much impaired.

The destruction of lipase through the gastric secretions has already been pointed out. When fed in the form of raw chopped pancreas some seems, however, to reach the small intestine in an uninjured state. It is a fact of clinical importance, therefore, that SANDMEYER has found that the absorption of fats in dogs deprived of their pancreas is much increased through the feeding of raw minced pancreas.–

CHAPTER XVII.

THE ALIMENTARY TRACT AS AN ABSORPTIVE SYSTEM.
(*Concluded*).

10. The **Absorption of Proteins.**—Since the proteins, unlike the salts and water, and like the fats and certain carbohydrates, suffer profound changes in their passage through the alimentary tract, the question arises first of all, Are the proteins absorbed entirely or in part as such, or do they first suffer a decomposition into simpler substances before they pass through the absorbing wall of the intestinal tract?

Physico-chemical reasons indicate that the proteins are not absorbed as such, or only in very small amounts, for belonging as they do to the general class of the colloids they are not able, practically speaking, to diffuse through a colloidal membrane such as the alimentary tract presents. Experimental facts agree entirely with this reasoning. As FRIEDLÄNDER'S [1] experiments have shown, acid-albumin does not disappear when introduced into a well-washed loop of intestine. In opposition to this are the experimental results of most other observers. BRÜCKE, VOIT, BAUER, CZERNY, LATSCHENBERGER, NEUMEISTER, and MUNK all believe that no inconsiderable amounts of such substances as egg-albumin, myosin, blood-serum, and acid-albumin can be absorbed as such from carefully cleaned loops of intestine in the course of one to four hours. The disappearance of even 10 or 20 percent of the protein introduced in this way does not prove necessarily that it was absorbed in the " native " state. The fact that the

---

[1] FRIEDLÄNDER, Zeitschr. f. Biol., 1896 XXXIII, p. 274.

amount of peptones in the blood leaving the intestines is not increased is an inconclusive argument, for they are not increased to any marked degree even under conditions when we know that the protein is being absorbed in the form of the soluble digestion products. From our knowledge of the almost universal distribution of proteolytic enzymes throughout the tissues and fluids of the animal body and their great activity in the living organism, it does not seem too hazardous to believe that even that portion of a protein solution which is believed to be absorbed in the "native" state is in reality absorbed in the form of peptones, or still more probably in the form of the simple crystalline products of proteolytic activity.

But even if we allow that proteins *may* be absorbed as such the amount that passes through the wall of the alimentary tract in this form must be small. . Just how small cannot, of course, be said, but the knowledge that adequate means exist in the body to split all the protein of an ordinary meal in the course of a few hours, and that the diffusion velocity of even such complex substances as the peptones is several times that of the proteins from which they are derived, indicates that under ordinary circumstances only a small fraction, if any at all, of the protein of an ordinary meal is absorbed in an unchanged state.

If protein is not absorbed as such, then in which of the various forms into which it is changed under the influence of the proteolytic ferments is it absorbed? This is a question which it is exceedingly difficult to answer, for the multitude of experiments that have been performed by many different investigators have not in any sense yielded results which are either entirely satisfactory or capable of only one interpretation. The burden of experimental evidence seems to indicate, however, that most, if not all, protein is absorbed in the form of the very simple crystalline products of proteolysis with which we became acquainted in the discussion of acid- and alkali-proteinase (pepsin and trypsin), and pro·

tease (erepsin). The arguments which point in this direction are the following: The proteoses still stand close to the so-called "typical" colloids, and their absorption as such is open to the same objections from a physico-chemical standpoint which were raised against the albumins from which they are derived. The same holds true of many of the peptones (in KÜHNE's sense), though some of them approximate the crystalloids in their physical behavior. In any case, both classes of substances are readily decomposed under the conditions existing in the body. As agencies bringing about such a decomposition we have the already-mentioned proteolytic ferments. Recent work indicates that acid-proteinase approximates at least qualitatively the proteolytic activity of alkali-proteinase, and in protease we have a newly discovered ferment which seems to be most energetic in breaking down proteoses and peptones into the simple crystalline products which have long been considered characteristic of the activity of alkali-proteinase (trypsin). Finally, if the proteins are really broken up into simple crystalline substances before being absorbed it ought to be possible to find them in the alimentary contents. This has been done by KUTSCHER and SEEMANN, who succeeded in obtaining leucin, tyrosin, and lysin from the intestinal contents of the dog. The amount which they obtained was not large, and this is often taken as an argument to show that not all or even a large part of the protein of a meal is split as far as the mono- and diamino-acids. The objection is perhaps scarcely a valid one, for not only are these substances readily diffusible so that a collection of a considerable amount is rendered well-nigh impossible, but all the proteolytic change does not occur in the lumen of the gut; part takes place in the wall of the intestine, where protease (erepsin) is present in large amounts and into which at least some of the peptones diffuse easily. All these facts seem to indicate, therefore, that the proteins are absorbed, at least in the main, in the form of the simplest digestion products.

Through what channel or channels are the proteins carried away from the alimentary tract? It has been found that the same holds true here as in the case of the carbohydrates. Under ordinary circumstances practically all the protein of a meal passes from the alimentary tract through the blood-vessels and only when excessive amounts are fed does a small percent pass over into the lymphatic circulation. SCHMIDT-MÜLHEIM showed this to be true when he found that the absorption of the proteins from the alimentary tract is not impeded when the main lymphatic channels coming from the alimentary tract are tied off. MUNK and ROSENSTEIN's [1] studies on a patient with a lymphatic fistula through which most of the visceral lymph was poured out externally show this in a still better way. When 80 to 103 grams of lean meat were fed this patient at a single meal (which more than covers the entire amount of protein consumed by the ordinary individual in a day) it was found that neither the total amount of lymph nor the percent of nitrogen in the lymph was markedly increased during the twelve hours following the meal. L. B. MENDEL [2] has come to similar conclusions from his experiments on dogs. Even when excessive amounts of protein are fed scarcely more than one-fifteenth of the whole amount that is absorbed passes from the alimentary tract through the lymphatics.

The relative importance of the stomach and of the intestine with its attached pancreas as organs concerned in the digestion and absorption of the proteins is somewhat difficult to determine, and the data we have at our disposal vary greatly. Since CZERNY, in 1878, removed the entire stomach of a dog and kept him in good health for six years afterward (when he was killed for study) no one has seriously questioned the statement that the stomach is not essential to good health, provided only a proper diet be observed. LUDWIG and OGATA

---

[1] MUNK and ROSENSTEIN: Ergebnisse d. Physiologie, 1902, I, 1te Abth., p. 312.

[2] L. B. MENDEL: American Jour. of Physiol., 1899, II, p. 137.

were able to show that the nitrogenous excretion of dogs could be covered entirely through the nitrogen obtained from finely minced meat, eggs, etc., introduced directly into the duodenum through a fistula, and more recently a number of operations carried out on human beings in which all or nearly all the stomach has been removed have shown that absence of this organ does not jeopardize life. The pancreas with the small intestine is able to take care of all the protein necessary for the life of the individual. For reasons which have already been pointed out, such patients do best on sterile food, for the gastric juice is no longer present to reduce the number of bacteria consumed with the ordinary food; and since an organ is no longer present which acts as a reservoir for the food and gives small amounts periodically to the intestine, repeated feedings only can be well tolerated. Since the acid of the gastric juice is best able of all the alimentary secretions to act upon the connective tissues, these appear in the fæces after the stomach is gone. ✓

The experiments which have been made to determine the importance of the stomach as an absorptive organ for the proteins are not free from criticism. They seem to indicate, however, that a small percent, perhaps, of the total amount of the protein in an ordinary meal may be absorbed by the gastric mucous membrane. As the chief absorptive organ cf the proteins we must regard the small intestine, more especially the upper half. This absorptive power decreases apparently from above downwards, reaching a very low grade in the large intestine.

According to most observers the pancreas plays a much heavier rôle than the stomach in the digestion of the proteins. Total extirpation of the pancreas is fatal, for this is followed by a diabetes which ends the life of its victim in a few days. For this reason it becomes necessary in experiments on the digestive functions of the pancreas to limit oneself to occlusion of the pancreatic duct or ducts by means of ligatures, or to only partial excision of the pancreas, or total excision with transplantation of a portion of the organ to some other part of the body. These various procedures have been adopted by different investigators. The results obtained vary greatly. All seem to agree, however, that the digestion and absorption of proteins are interfered with markedly when the pancreatic juice no longer pours into the duodenum. This is not surprising when it is remembered how well protein digestion may go on after removal of the stomach. Generally speaking, less than half the protein of an ordinary meal is digested when the pancreas is gone. HARLEY found only 18 percent digested after removal of the pancreas, though DEUCHER found 80 percent utilized after complete occlusion of the pancreatic duct.[1] The proteolytic activity must evidently be sought in these cases in the secretion of the stomach, augmented by those found in the secretions and mucous membrane of the small intestine.

What, now, is the fate of the protein which disappears from the lumen of the alimentary tract? Of fundamental importance is the fact that even after a meal rich in proteins neither peptones nor proteoses appear in either the blood or the lymph. This has been shown to be true in various ways: Direct analysis of the blood leaving the intestine points in this direction. Furthermore, peptones injected directly into the blood are excreted through the kidneys, and normally no peptones can be found in the urine, even after a very heavy protein meal. The peptones lower the blood pressure when injected into the general circulation, yet a lowered blood pressure is never observed after the consumption of large amounts of protein, or even after the introduction of large amounts of peptone into the alimentary tract. All this indicates clearly that peptones do not get into the blood or lymph (at least in sufficient amounts to be recognized by the analytical means ordinarily employed). Nor do peptones and proteoses in the blood issuing immediately from the in-

---

[1] See MUNK: l. c., p. 316.

testine appear to be caught and held by some organ, such as the liver or spleen, before passing over into the general circulation. This has been shown by comparative analyses of the blood obtained from the portal vein, for example, and that obtained from a large artery, as well as by the experiments already cited.

What, then, does become of the proteins which enter the alimentary tract? It is clear that if they disappear from the lumen of the intestine and cannot be discovered in the blood leaving the alimentary tract as proteoses and peptones they must exist here in some other form. In what form cannot be easily said; though there are two possibilities, either or both of which may be correct, and both of which have experimental foundation.

The first of these is that the proteins, be these absorbed in whatever way we may consider the correct one, are reconstructed in their passage through the epithelium of the stomach or intestine into complex albumins and globulins which we are unable to distinguish from those found normally in the blood. This means that if we believe all the protein of a meal to be absorbed in the form of KÜHNE's peptones, or perhaps in the form of mono- and diaminoacids, that these are reconverted into complex albumins in passing through the alimentary wall.

The experiments of ABDERHALDEN and SAMUELY [1] are of great interest in this connection. These show that the composition of the blood, so far as its protein constituents are concerned, is apparently entirely independent of the character of the proteins consumed by the individual. Analysis of six litres of blood removed from the veins of a horse showed that it contained protein bodies which yielded after hydrolysis:

Tyrosin.................... 2.43 percent
Glutamic acid .... 8.85 "

[1]. ABDERHALDEN and SAMUELY: Zeitschr. f. physiol. Chem., 1905, XLVI, p. 193.

After starvation for a week, in order to free the alimentary
tract from food, analysis of another six litres of blood yielded
practically the same results:

Tyrosin..................... 2.60 percent
Glutamic acid .... 8.20   "

The horse was now fed a protein (gliadin) which is much
richer in glutamic acid than blood serum and contains about
the same percent of tyrosin.   Even when 1500 to 2500 grams
of the protein were fed analysis of the blood showed practically no variation in its composition.   The following table
gives the results of three experiments:

	1500 gms. Percent.	1500 gms. Percent.	2500 gms. Percent.
Tyrosin.	2.24	2.52	2.48
Glutamic acid	7.88	8.25	8.00

It seems, therefore, as though an actual synthesis of protein occurs in the wall of the intestine itself.   From the
most varied kinds of protein consumed by an animal are constructed first of all the serum albumin and serum globulin found
in the blood-serum of that animal, and from this serum, so
constant in its composition, the different cells of the body each
take the substances necessary for their purposes.   Indirectly
we find in these considerations support for the idea that the
proteins are absorbed from the alimentary tract only in the
form of the simple digestion products, for these can most
readily be joined chemically to produce the protein bodies
characteristic of the animal under consideration.

As ABDERHALDEN [1] has well pointed out, these conceptions
place the functions of the alimentary tract and its enzymes in
an entirely new light.   They together guarantee a proper metabolism of the body as a whole.   The ferments through their
action on the various foodstuffs break these up into their
elements, which are the same even when derived from the

---

[1] ABDERHALDEN: Lehrbuch d. physiol. Chem., Berlin, 1906, p. 232.

most varied parent bodies. The alimentary tract then synthesizes from these elements the complex substances found in the blood. In disturbances in general metabolism involving the alimentary tract it is not so much the impaired absorption that is so important as the impaired assimilation. That this synthesis of protein plays an important rôle in the metabolism of the animal has been shown very strikingly by ABDERHALDEN and RONA.[1] These authors were able to keep animals in nitrogenous equilibrium by feeding them, in addition to an ordinary mixture of fats and carbohydrates, casein which had been previously digested through alkaliproteinase, so that 80 to 85 percent of it consisted of aminoacids mixed with simple polypeptides, the remaining 15 to 20 percent of more complex polypeptides, but not sufficiently so to give a biuret reaction.

As agencies capable of such a reconstruction of the albumin molecule have been mentioned the chemical forces of the epithelial cells themselves and the activities of JULIA BRINCK's *micrococcus restituens*. That a microörganism living upon the surface of the alimentary mucous membrane should be of any importance to its host by being able to reconstruct from the products of proteolysis the protein itself (which would again have to be broken up before it could' pass through the epithelial cells in any amount) is, no doubt, too improbable to need serious consideration. There is much more likelihood that the chemical forces residing in the epithelial cells are capable of reconstructing albumins from the products of proteolysis. Elsewhere[2] experiments have been discussed which seem to indicate that the activity of the proteolytic ferments is reversible. Were this true, then the synthesis of albumins within the epithelial cells lining the alimentary tract, in a way entirely analogous to the synthesis of fat in this locality under the influence of lipase,

---

[1] ABDERHALDEN and RONA: Zeitschr. f. physiol. Chem., 1904, XLII, p. 528; ibid., 1905, XLIV, p. 108.

[2] See p. 140.

might readily be explained. The presence of such ferments as alkali-proteinase (trypsin) has been demonstrated not only in these cells but in practically every cell and fluid of the body.

A second reason why the products of proteolysis—proteoses, peptones, or amino-acids—do not appear in the blood leaving the intestine may reside in the fact that they are in their passage through the intestinal wall broken into yet simpler substances. The proteinase and protease present in the cells of the mucosa may well continue their work until none of the higher polypeptides are left. But, even if all the protein is first split into the simple mono- and diamino-acids before passing into the blood, as seems to be the case, all these acids need not necessarily go to build up albumins and globulins once more. They may be broken into yet simpler substances. There exist in the mucous membrane of the alimentary tract and in various organs in the body ferments which are capable of bringing about such a destruction of mono- and diamino-acids. With one such ferment we are acquainted, through KOSSEL and DAKIN'S work on arginase, which is a ferment capable of acting upon arginin and splitting this into urea and ornithin. The discovery of this ferment, as also the well-known fact that the excretion of urea in the urine reaches its maximum shortly after the consumption of a protein meal, seems to indicate that a portion, at least, of the albuminous bodies which constitute our food are rapidly broken up into the *ultimate* products of protein metabolism. Apparently, therefore, a part of the protein we absorb goes to maintain the proportion of albumin and globulin found in the blood, and so is distributed to the cells and tissues of the body, while another serves as a source of energy and leaves the body in the form of urea.

This is perhaps the best place in which to discuss the physiological importance of protease (erepsin) as a digestive enzyme, and of importance, in consequence, in the absorption of the proteins. It is evident that from the nature of its action alone it might well play a rôle not inferior to that of acid- or alkali-proteinase (pepsin or trypsin). Not only does protease act on casein directly, but it is able to bring about the split- ting of proteoses and peptones as energetically and as rapidly as either acid- or alkali-proteinase. In fact, it need not surprise us if in the near future we discover that some of the cleavages of protein which we have thus far attributed to alkali-proteinase are in reality brought about through the protease existing beside the proteinase as an impurity.

Protease is found not only in the secretions of the small intestine but also in the cells of the intestinal mucosa. Since much more is found in the latter location than in the former, it may well be concluded that the protease within the cells is of greater physiological importance than that contained in the secretions. Self-apparent also is the fact that the power of diffusion possessed by the proteoses in part, by the peptones (in KÜHNE's sense) in much larger part, plays an important rôle in rendering the protease found in the cells an active agent in the demolition of the protein molecule.

The following experiment [1] illustrates the rapidity with which a peptone solution disappears from the intestine. A loop of intestine 50 cm. long is carefully taken out of the abdominal cavity of a dog and after being ligatured at both ends is opened and well washed. 39 c.c. of a peptone solution containing 0.49 gm. nitrogen are introduced into the intestine and the loop replaced in the abdominal cavity. At the end of an hour the loop is again taken out and the amount of unabsorbed peptone solution determined. It is found that 0.296 gm. nitrogen has been absorbed, in other words, over 60 percent.

When the 95 cm. of intestine which were not used in this experiment are taken into consideration, simple calculation shows that this dog would have been able to absorb from its entire small intestine more than 0.8 gm. nitrogen per hour.

---

[1] COHNHEIM: Zeitschr. f. physiol. Chemie, 1902, XXXVI, p. 13.

This figure, which is by no means the highest that might be given for experiments of this sort, indicates how rapidly peptones disappear from the small intestine, especially when it is added that at this rate a dog could absorb in one hour nearly enough nitrogenous material to suffice for its daily existence.

It remains for us to show that this rapid absorption of peptones is determined by the action of protease upon them, whereby they are dissociated into the very diffusible crystalline digestion-products which have been enumerated above; and that this dissociation is not brought about by such a ferment as alkali-proteinase (trypsin). While under ordinary circumstances both alkali- and acid-proteinase are active in bringing about the dissociation of the protein which we take in with our food, under the experimental conditions outlined above this is not the case, for a well-washed loop of intestine contains no alkali-proteinase (trypsin), or at best only very little. This has been repeatedly shown by experiments in which proteins, such as fibrin, are introduced directly into the small intestine, when it has been found that they disappear very slowly (in days or even weeks).

CHAPTER XVIII.

THE ALIMENTARY TRACT AS AN EXCRETORY SYSTEM.

1. General Clinical and Experimental Considerations.— Practical medicine has for a long time recognized empirically the function of the alimentary tract as an organ of excretion. As an example we need only cite the practice of purgation in cases of renal disease, in which the effort is made to carry off the poisonous metabolic products of the body through the intestinal tract. The beneficial effects following this therapeutic procedure are illustrated in the clinical experiences of every day.

The scientific basis of this practice has not as yet been definitely established through experiment. It is true that there exist a large number of isolated facts, which show that various poisons, no matter how introduced into the living organism, are eliminated through the alimentary tract. Every physician is acquainted with the elimination of the iodides through the salivary glands after introduction of these salts into the stomach;[1] with the elimination of arsenic and morphin[2] through the stomach, even when these substances do not enter the organism through the intestinal tract, and with the excretion of iron compounds through the small intestine. MENDEL and THACHER[3] have recently collected a

---

[1] PENZOLDT and FABER: Berliner klin. Wochenschr., 1882, XIX, p. 363.

[2] KUNKEL: Handbuch d. Toxikologie, 1901, p. 54.

[3] MENDEL and THACHER: American Journal of Physiology, 1904, XI, p. 5.

large number of instances illustrating the fact that various organic and inorganic substances are eliminated through the alimentary tract. For the most part, however, quantitative determinations, showing the amounts of the various substances eliminated in this way and the amounts cast off through the other emunctories of the body, for instance, the kidneys and skin, have not been made in these earlier experimental studies.

With the especial purpose of determining the relative importance of the kidneys and alimentary tract as organs of excretion for certain of the inorganic compounds, MENDEL and his pupils, HANFORD and THACHER, have taken up this problem.[1]

MENDEL and THACHER used strontium (in the form of strontium acetate) in their experimental studies, as this is an element which does not occur normally in the ordinary laboratory animals, and hence can be readily recognized in the different tissues spectroscopically. As the experiments were designed to determine the function of the alimentary tract as an excretory organ the salt could not be given by mouth. A four percent solution of strontium acetate was in consequence injected subcutaneously with all aseptic precautions; or at times intraperitoneally or intravenously. The animals were kept in metallic cages and the urine and fæces collected separately. A series of experiments performed on dogs, cats, and rabbits yielded the following interesting results.

Strontium is eliminated only to a small extent through the kidneys, even when this element is introduced in the form of a salt directly into the circulation. The excretion in the urine begins shortly after the injection and usually ceases within twenty-four hours. By far the larger portion of the strontium is excreted through the fæces, and it is immaterial how the element has been given. The place of excretion is

---

[1] HANFORD: American Journal of Physiology, 1903, IX, p. 235; MENDEL and THACHER: ibid., 1904, XI, p. 7.

apparently limited to the region of the alimentary tract beyond the stomach.

The following experiments will serve to illustrate the foregoing. As strontium tends to be stored in certain tissues of the body and is only eliminated slowly, the dejecta of the animals experimented upon have to be studied for several days. A dog weighing 6 kilos received a number of subcutaneous strontium acetate injections, the total amount of strontium injected being 0.543 gm. The urine and fæces were collected separately for 21 days. The total urine contained an unweighable trace of strontium. The total fæces contained 0.0998 gm. of this element.

At times, however, a much larger percentage of the injected strontium can be recovered. A dog weighing 14 kilos was given 0.28 gm. strontium (in the form of the acetate) subcutaneously. At no time for 17 days subsequently could strontium be detected in the urine. From the fæces 0.237 gm. were recovered during this period.[1]

In a series of experiments which Professor MENDEL [2] has communicated to me by letter, he has been able to show that barium behaves not unlike strontium. If barium chloride is introduced into the body even in other ways than through the mouth, it is eliminated almost exclusively by the intestinal tract. After the first twenty-four or forty-eight hours the urine shows no signs of containing the element barium, even though the fæces contain the substance for days afterward. Barium tends to be stored in the body even more readily than strontium and so is eliminated more slowly.

Rubidium belongs in the class with sodium, and like this leaves the body chiefly through the kidneys. To a certain extent, however, rubidium also is eliminated through the intestinal tract.

---

[1] MENDEL and THACHER: American Journal of Physiology, 1904, XI p. 14.

[2] Personal letter dated New Haven, Connecticut, July 23, 1905.

The question of the excretory function of the alimentary tract has also been studied by J. B. MacCallum.[1] This observer has confirmed the experiments of some of the older students of œdema that sodium-chloride solutions of the osmotic concentration of the blood, or somewhat higher, when injected into the circulation of rabbits, cause a greatly increased secretion of fluid into the intestinal tract. The amount of fluid eliminated in this way varies both with the rate and the quantity of the salt solution that is injected. In one experiment in which 500 c.c. of salt solution were injected intravenously, 14.46 percent were eliminated through the intestinal tract. In another experiment 9 percent of the total quantity of fluid injected was eliminated in this way, and in a third, 10.25 percent. When the kidneys are removed the amount excreted through the intestine is somewhat higher—16.6 percent in one experiment. The intestinal tract behaves therefore not unlike the kidneys under similar circumstances. We are well acquainted with the effect of intravenous injections of salt solutions in increasing the output of urine. The increased secretion from the intestinal tract is therefore not unlike the polyuria brought about by similar means.

Interestingly enough, the other salts which bring about an increased secretion of urine also bring about in a similar way an increased secretion from the intestine. Experiment has shown that the diuretic salts and the saline cathartics are the same. So far as the excretion of water from the body is concerned, therefore, we may well look upon the intestinal tract as supplementary to the kidneys.

MacCallum made no determinations of the relative amounts of the various salts which are eliminated through the kidneys and intestine. It is clear from Mendel's experiments, however, that these must differ with the different elements. The

---

[1] MacCallum: University of California Publications, Physiology, 1904, I, p. 125.

former has, however, found in his own experiments and has collected facts from the literature which show that substances which are ordinarily believed to be excreted chiefly or exclusively through the kidneys are eliminated also by the intestine. CLAUDE BERNARD found as far back as 1859 that urea occurs in the saliva and the gastric juice, and that in nephritis and after removal of the kidneys this substance is excreted to a large extent by the intestine. From this fact CLAUDE BERNARD concludes that the intestine may supplement the kidneys in eliminating the various constituents of the urine. The presence of urea in the intestinal juice of sheep has been demonstrated by PREGL, and MacCALLUM found this substance normally present in the same secretion of the rabbit.

Of great interest is the fact discovered by MacCALLUM that in a certain form of experimental diabetes the sugar is eliminated, not only by the kidneys but also through the intestinal tract. As first shown by BOCK and HOFFMANN and confirmed by KÜLZ's and my own experiments, the intravenous injection of rabbits with a sodium chloride solution of a somewhat higher concentration than the sodium chloride in the blood brings about a transient excretion of glucose in the urine.[1] The observations of MacCALLUM show that under these conditions sugar is eliminated also by the stomach and intestine and apparently in about the same concentration as in the urine. The excretion of the carbohydrate through the intestinal tract can be made more apparent by removing the kidneys, but even when they are left intact, sugar is nevertheless excreted to some extent through the alimentary tract. How great a rôle this route of excretion plays in the ordinary cases of human diabetes is an interesting question which has not as yet been investigated.[2]

---

[1] MARTIN H. FISCHER: University of California Publications, Physiology, 1903, I, pp. 77 and 87; PFLÜGER's Archiv., 1904, CVI, p. 80; ibid., 1905, CIX, p. 1.

[2] My analysis of the fæces in two cases of human diabetes failed

The above facts, to which many more could be added from medical literature, begin to give us an experimental foundation for the long-established empirical practice of utilizing the intestinal tract as an organ of excretion in those diseases in which the function of the kidneys is impaired. Not only has it been shown that for many substances the intestinal tract is the chief excretory organ, but also that the intestine behaves in many respects not unlike the kidneys. The same salts which act as diuretics act as cathartics, and sugar and urea, which by many have been looked upon as substances excreted only by the kidneys, may be lost from the body through the intestinal tract also.

The problems which suggest themselves in this domain of the excretory function of the alimentary tract, so often touched upon but so little studied, are many. In how far can we substitute the activity of the intestinal tract as an excretory organ for that of the kidneys in ridding the organism of the various substances found in the urine both in health and disease? Each of the substances found in this excretion, from the ordinary salts to the most complex organic poisons, such as the toxins, will have to be investigated with this problem in mind, and the results obtained may well be ex-

---

to show sugar in sufficient amounts to be recognized by ordinary laboratory methods even when 600 to 700 grams of sugar were being excreted in the urine. According to experiments carried on with GERTRUDE MOORE, practically no sugar escapes through the gastro-intestinal tract in rabbits rendered diabetic through puncture of the medulla. But this happens in such rabbits as soon as a sodium chloride solution is injected intravenously. A sodium chloride solution too dilute to bring about a glycosuria by itself is able to do this. The sodium chloride therefore renders the gastro-intestinal mucosa permeable in one direction to a substance to which it was formerly impermeable. An analysis of MacCALLUM's experiments shows that he never injected sodium chloride in sufficient amounts to bring about a glycosuria. He rendered his rabbits diabetic through the use of morphin, and obtained sugar in the gastro-intestinal secretions through subsequent intravenous injections of sodium chloride solutions.

pected to be of the utmost practical value in medicine. It was shown above to how great an extent certain inorganic substances are eliminated by the alimentary tract. FLEXNER has found that the toxin of the SHIGA bacillus, no matter how introduced into the body, is eliminated through the intestinal tract, and in this elimination gives rise to the ulcers found in the dysentery produced by this organism. The recognition of the fact that morphin is secreted into the stomach, even when subcutaneously introduced, has led to the practice of gastric lavage for morphin poisoning; and the knowledge that arsenic appears in the gastro-intestinal tract, no matter how it is given, has led to scientific methods of recognizing cases of poisoning by this element, not only for diagnostic but also for medico-legal purposes.

2. The Character of the Alimentary Contents. The Fæces. —The alimentary contents differ markedly in general appearance and chemical composition in the various portions of the alimentary tract. The reasons for this are readily apparent. Not only does the food of different animals differ, but it is not always the same in the same animal. In its passage from the mouth to the anus it is acted upon by the various portions of the canal through which it passes, and changes are wrought in it of a mechanical and chemical character. Not only is its physical state of aggregation progressively altered in this way, but it has poured out upon it, one after the other, a number of secretions, each of which, by virtue of the substances contained in it, alters the chemical composition of the alimentary contents. The metabolic products excreted by the bacteria accomplish a similar end. To the changes brought about in this way must be added those induced through the absorption from the alimentary contents of certain of the substances contained in them. All these are together responsible for the progressive change which the food suffers in its passage from the mouth to the anus.

The food enters the mouth in a state of coarse division,

and through the action of the teeth is more finely divided. With most individuals the food as it slips into the stomach still contains large masses of undivided food, made up more particularly of pieces of meat and smaller but still coarse pieces of boiled potato, bread, etc. The mechanical action of the stomach, together with the action of the large quantities of gastric juice poured out under physiological conditions upon the food in this locality, serves to change even the external appearance of the mixed food. Not only are the larger pieces of starchy food broken into smaller ones through the muscular contractions passing over the stomach, but pieces of swallowed meat suffer similar changes in their state of aggregation. The connective tissue found in them is acted upon by the gastric juice and the cellular elements in consequence are allowed to fall apart. Fatty tissues suffer a similar change, and the fat contained within the cells becomes free. If any fats have been consumed which at ordinary temperatures are solid, but melt at body temperatures, these melt. The formation of peptones from the protein of the meal imparts to the gastric contents a bitter taste.

The acid, liquid, partially digested gastric contents pass over in small amounts into the duodenum. Here they have poured out upon them, as soon as they pass the pancreatic duct, the bile and pancreatic juice. The color of the bile imparts itself to the alimentary contents, and the alkaline reaction of the pancreatic juice reduces the acidity of the food as it has escaped from the stomach. Throughout the small intestine the food exists only as a sticky, mucinous, brownish-yellow mass which adheres more or less closely to the mucous membrane of the alimentary tract. The viscosity of this mass increases gradually from above downward, and nowhere throughout the small or large intestine do the alimentary contents show to the naked eye any of the characteristics of the food originally consumed by the individual. These are quite effectually lost even in the stomach after digestion has gone on for two or three hours. The yellowish-brown contents of the small intestine assume a somewhat darker color when the ileocæcal valve is passed, and from here down to the anus the alimentary contents lose water, and in consequence increase in consistency very rapidly. Semisolid scybala begin to appear in the transverse and descending portions of the large intestine, and in the lowermost portions of the large bowel the fæces proper are formed. Throughout the small intestine the alimentary contents have no fæcal odor under ordinary circumstances. This is developed after the ileocæcal valve is passed.

The question of the *reaction of the gastro-intestinal contents* has within the last few years come up again for discussion. In consequence of more careful analyses, made possible through advances in our knowledge of indicators, some of our older ideas regarding the reaction of the intestinal contents will have to be set aside and newer ones adopted. The reaction of the gastric contents after these have remained in the stomach for some time is, under normal circumstances, acid, and this fact has never been disputed since the middle of the last century. Broadly speaking, the contents of the small intestine have always been considered alkaline. Except for the first few inches of duodenum in which the fresh acid gastric contents may be found, the contents of the remaining portion of the alimentary tract were considered alkaline in reaction, because they had poured out upon them the "pronouncedly alkaline" secretions of the pancreas, liver, and intestine itself. Neglecting for the time being the reaction of these juices, let us ask what means were employed to ascertain the reaction of the gastro-intestinal contents? For the most part only one indicator was used to determine their acidity or alkalinity, litmus, and this in its poorest form, as litmus paper. Litmus in any form, however, is an exceedingly fallacious indicator, at times even useless, in solutions in which carbonates are present, for litmus is not sufficiently sensitive toward carbonates. For this reason litmus may even show an alkaline reaction in solutions which are really neutral or even acid.

A number of authors have in recent years employed indicators other than litmus (such as lacmoid, phenolphthalein, methyl orange, alkanna, rosolic acid, alizarin, curcuma and trapæolin) to ascertain the reaction of the intestinal contents. MUNK,[1] who experimented in this way on dogs and hogs fed on various diets after a period of fasting—diets predominantly protein, or protein and fatty, or mixed in character—comes to the following conclusions: The contents of the duodenum, jejunum, or ileum, in carnivorous as well as in omnivorous animals, no matter how fed, at no time show an alkaline reaction, if only indicators sufficiently sensitive to indicate the presence of carbonates and fatty acids be employed. On a pure meat diet the duodenal and upper jejunal contents are distinctly even though only faintly acid. Beyond this point and to the ileocæcal valve the intestinal contents are neither definitely acid nor alkaline, and may, in consequence, be considered neutral. As soon as fat is added to the diet, that is, protein and fat are fed together, the contents of the entire small intestine show an acid reaction, attributable, no doubt, in the main to the presence of free fatty acids formed through the action of lipase upon the neutral fats. In fact, under no circumstances, even when large amounts of carbohydrates are given together with protein, do the contents of the jejunum or ileum ever show an alkaline reaction.

The experimental observations of MUNK are corroborated by clinical findings on human beings. MACFADYEN, NENCKI, and SIEBER,[2] who observed two cases of fistula of the ileum just above the ileocæcal valve, found that the contents of the alimentary tract when they passed this point were always acid in reaction when the patients were given a mixed diet.

---

[1] MUNK: Centralblatt für Physiologie, 1902, XVI, p. 33.

[2] MACFADYEN, NENCKI, and SIEBER: Archiv für experimentelle Pathologie und Pharmakologie, 1891, XXVIII, p. 311.

The acidity was due to organic acids. These acids no doubt inhibit the development of those bacteria which are capable of producing putrefaction of the protein substances present in the small intestine. For this reason no protein putrefaction occurs here under normal circumstances, or only rarely.

If what has been said above concerning the reaction of the intestinal contents is true, then how are we to reconcile it with the observations that have been made on various ferments and their behavior toward acids and alkalies? Alkaliproteinase, for example, is generally stated to act best in an alkaline medium. In the body of various animals, including man, this ferment must however work for the most part in an acid, at the best in a neutral medium. Evidently, therefore, alkali-proteinase does not work under the most favorable conditions, so far as reaction of surrounding medium is concerned, in the body; or else the statements of the various students of the effect of acids and alkalies on alkali-proteinase must be revised. So far as the former of these possibilities is concerned, we know that alkali-proteinase will act in neutral or acid media, even very powerfully in neutral media. So far as the latter possibility is concerned, it is entirely probable that the use of indicators better adapted to the investigation of the problem will show that the alkalinity in which alkali-proteinase acts best is lower than we formerly supposed, and that some of the solutions in which alkali-proteinase is known to act very well and which are ordinarily stated to be alkaline are really neutral or even acid in reaction. Under normal circumstances, therefore, alkali-proteinase may still be considered as acting in the intestine under only slightly unfavorable circumstances.

What has been said of alkali-proteinase holds also for the other enzymes. Unquestionably the acid reaction of the intestinal contents may be of service to certain of the ferments, such, for example, as the acid-proteinase of the stomach and duodenum. Finally, even if the acid reaction of the contents of the small intestine reduces the activity of the digestive ferments found here, it reduces still more markedly that of the ferments found in the bacteria, for the growth of these is decidedly kept in check throughout this hollow viscus, and their putrefactive action upon the food (which, under pathological conditions, may become of a serious character) prevented to a large extent.

The amount of fæces cast off by an animal is dependent upon the amount and character of the food consumed. Other things being equal, a large amount of food will yield a greater amount of excrementitious material than a smaller one. The extent to which a food can be absorbed is however of great importance. Fine white-flour breads are, for example, absorbed more perfectly than rye or whole-wheat breads, because they represent more nearly pure starch in a form that can be acted upon by the digestive ferments. Mashed potatoes yield less fæcal matter than boiled potatoes, for in the former the cellulose membranes surrounding the cells are more perfectly broken than in the latter, and the starch absorption is in consequence more perfect. Because of the large amount of cellulose they contain, the vegetable diets in general yield a larger amount of fæcal matter than meat diets. The fæces do not, however, represent only remnants of food that cannot be digested, but also a certain amount that can be, but, for various reasons, has not been digested. In addition it must not be forgotten that the secretions of the gastro-intestinal tract itself and of the glands connected with it contribute largely toward making up the body of the fæces.

The color of the fæces is variable. A meat diet yields dark, one of bread, lighter fæces. The color is determined in large part by the bile, being gray in its absence, yellowish or yellowish brown in its presence. For the most part, it is not the bile pigments themselves that give this color, but certain of their derivatives formed after the bile has escaped into the intestine.

INDEX OF AUTHORS.

323

INDEX OF SUBJECTS.

333

Short-Title Catalogue

OF THE

PUBLICATIONS

OF

JOHN WILEY & SONS

New York

London: CHAPMAN & HALL, Limited

ARRANGED UNDER SUBJECTS

Descriptive circulars sent on application.   Books marked with an asterisk (*) are sold at *net* prices only.   All books are bound in cloth unless otherwise stated.

AGRICULTURE—HORTICULTURE—FORESTRY.

Armsby's Principles of Animal Nutrition............................8vo, $4 00
Budd and Hansen's American Horticultural Manual:
    Part I. Propagation, Culture, and Improvement...............12mo, 1 50
    Part II. Systematic Pomology................................12mo, 1 50
Elliott's Engineering for Land Drainage...........................12mo, 1 50
    Practical Farm Drainage.   (Second Edition, Rewritten)........12mo, 1 50
Graves's Forest Mensuration.......................................8vo, 4 00
Green's Principles of American Forestry...........................12mo, 1 50
Grotenfelt's Principles of Modern Dairy Practice.   (Woll.)..........12mo, 2 00
* Herrick's Denatured or Industrial Alcohol........................8vo, 4 00
Kemp and Waugh's Landscape Gardening.   (New Edition, Rewritten.   In
    Preparation).
* McKay and Larsen's Principles and Practice of Butter-making.....8vo, 1 50
Maynard's Landscape Gardening as Applied to Home Decoration......12mo, 1 50
Sanderson's Insects Injurious to Staple Crops......................12mo, 1 50
Sanderson and Headlee's Insects Injurious to Garden Crops.   (In Prep-
    aration).
* Schwarz's Longleaf Pine in Virgin Forests........................12mo, 1 25
Stockbridge's Rocks and Soils.....................................8vo, 2 50
Winton's Microscopy of Vegetable Foods...........................8vo, 7 50
Woll's Handbook for Farmers and Dairymen........................16mo, 1 50

ARCHITECTURE.

Baldwin's Steam Heating for Buildings............................12mo, 2 50
Berg's Buildings and Structures of American Railroads...............4to, 5 00
Birkmire's Architectural Iron and Steel.............................8vo, 3 50
    Compound Riveted Girders as Applied in Buildings................8vo, 2 00
    Planning and Construction of American Theatres..................8vo, 3 00
    Planning and Construction of High Office Buildings..............8vo, 3 50
    Skeleton Construction in Buildings...............................8vo, 3 00

Briggs's Modern American School Buildings..........................8vo, $4 00
Byrne's Inspection of Materials and Wormanship Employed in Construction.
                   16mo, 3 00
Carpenter's Heating and Ventilating of Buildings.....................8vo, 4 00
* Corthell's Allowable Pressure on Deep Foundations.................12mo, 1 25
Freitag's Architectural Engineering.....................................8vo, 3 50
 Fireproofing of Steel Buildings...................................8vo, 2 50
Gerhard's Guide to Sanitary Inspections. (Fourth Edition, Entirely Re-
  vised and Enlarged).......................................12mo, 1 50
 * Modern Baths and Bath Houses...............................8vo, 3 00
 Sanitation of Public Buildings................................12mo, 1 50
 Theatre Fires and Panics 12mo, 1 50
Johnson's Statics by Algebraic and Graphic Methods.................8vo, 2 00
Kellaway's How to Lay Out Suburban Home Grounds...............8vo, 2 00
Kidder's Architects' and Builders' Pocket-book.................16mo, mor., 5 00
Merrill's Stones for Building and Decoration........................8vo, 5 00
Monckton's Stair-building..4to, 4 00
Patton's Practical Treatise on Foundations..........................8vo, 5 00
Peabody's Naval Architecture.......................................8vo, 7 50
Rice's Concrete-block Manufacture..................................8vo, 2 00
Richey's Handbook for Superintendents of Construction .... 16mo, mor. 4 00
 Building Foreman's Pocket Book and Ready Reference. . 16mo, mor. 5 00
 * Building Mechanics' Ready Reference Series:
  * Carpenters' and Woodworkers' Edition............16mo, mor. 1 50
  * Cement Workers' and Plasterers' Edition...........16mo, mor. 1 50
  * Plumbers', Steam-Fitters', and Tinners' Edition...16mo, mor. 1 50
  * Stone- and Brick-masons' Edition................16mo, mor. 1 50
Sabin's House Painting..12mo, 1 00
Siebert and Biggin's Modern Stone-cutting and Masonry.............8vo, 1 50
Snow's Principal Species of Wood.................................8vo, 3 50
Towne's Locks and Builders' Hardware........................16mo, mor. 3 00
Wait's Engineering and Architectural Jurisprudence..................8vo, 6 00
                  Sheep, 6 50
 Law of Contracts..8vo, 3 00
 Law of Operations Preliminary to Construction in Engineering and Archi-
  tecture..8vo, 5 00
                  Sheep, 5 50
Wilson's Air Conditioning...12mo, 1 50
Worcester and Atkinson's Small Hospitals, Establishment and Maintenance,
 Suggestions for Hospital Architecture, with Plans for a Small Hospital.
                 12mo, 1 25

ARMY AND NAVY.

Bernadou's Smokeless Powder, Nitro-cellulose, and the Theory of the Cellulose
 Molecule..12mo, 2 50
Chase's Art of Pattern Making.....................................12mo, 2 50
 Screw Propellers and Marine Propulsion........................8vo, 3 00
* Cloke's Enlisted Specialists' Examiner............................8vo, 2 00
 * Gunner's Examiner...8vo, 1 50
Craig's Azimuth..4to, 3 50
Crehore and Squier's Polarizing Photo-chronograph..................8vo, 3 00
* Davis's Elements of Law..8vo, 2 50
 * Treatise on the Military Law of United States.................8vo, 7 00
DeBrack's Cavalry Outpost Duties. (Carr.)...................24mo, mor. 2 00
* Dudley's Military Law and the Procedure of Courts-martial...Large 12mo, 2 50
Durand's Resistance and Propulsion of Ships........................8vo, 5 00
* Dyer's Handbook of Light Artillery..............................12mo, 3 00
Eissler's Modern High Explosives..................................8vo, 4 00
* Fiebeger's Text-book on Field Fortification..................Large 12mo, 2 00
Hamilton and Bond's The Gunner's Catechism......................18mo, 1 00
* Hoff's Elementary Naval Tactics.................................8vo, 1 50
Ingalls's Handbook of Problems in Direct Fire......................8vo, 4 00
* Lissak's Ordnance and Gunnery.................................8vo, 6 00

Ingalls's Handbook of Problems in Direct Fire 8vo, $4 00
* Lissak's Ordnance and Gunnery...................................8vo, 6 00
* Ludlow's Logarithmic and Trigonometric Tables:.................8vo, 1 00
* Lyons's Treatise on Electromagnetic Phenomena. Vols I. and II..8vo,each, 6 00
* Mahan's Permanent Fortifications. (Mercur.) ....8vo, half mor. 7 50
Manual for Courts-martial..16mo,mor. 1 50
* Mercur's Attack of Fortified Places............................12mo, 2 00
    * Elements of the Art of War................................8vo, 4 00
Nixon's Adjutants' Manual..24mo, 1 00
Peabody's Naval Architecture.....................................8vo, 7 50
* Phelps's Practical Marine Surveying............................8vo, 2 50
Putnam's Nautical Charts...8vo, 2 00
Rust's Ex-meridian Altitude, Azimuth and Star-Finding Tables........8vo, 5 00
* Selkirk's Catechism of Manual of Guard Duty...................24mo, 50
Sharpe's Art of Subsisting Armies in War. ....18mo, mor. 1 50
Taylor's Speed and Power of Ships   (In Press.)
* Tupes and Poole's Manual of Bayonet Exercises and Musketry Fencing.
                                                24mo, leather, 50
* Weaver's Military Explosives 8vo, 3 00
* Woodhull's Military Hygiene for Officers of the Line........Large 12mo, 1 50

ASSAYING.

Betts's Lead Refining by Electrolysis 8vo, 4 00
Fletcher's Practical Instructions in Quantitative Assaying with the Blowpipe.
                                                16mo, mor. 1 50
Furman and Pardoe's Manual of Practical Assaying. (Sixth Edition, Re-
        vised and Enlarged.).....................................8vo, 3 00
Lodge's Notes on Assaying and Metallurgical Laboratory Experiments .8vo, 3 00
Low's Technical Methods of Ore Analysis...........................8vo, 3 00
Miller's Cyanide Process...12mo, 1 00
    Manual of Assaying..12mo, 1 00
Minet's Production of Aluminum and its Industrial Use. (Waldo.)...12mo, 2 50
Ricketts and Miller's Notes on Assaying..........................8vo, 3 00
Robine and Lenglen's Cyanide Industry. (Le Clerc.)................8vo, 4 00
* Seamon's Manual for Assayers and Chemists.. ....Large 12mo, 2 50
Ulke's Modern Electrolytic Copper Refining.......................8vo, 3 00
Wilson's Chlorination Process....................................12mo, 1 50
    Cyanide Processes...12mo, 1 50

ASTRONOMY.

Comstock's Field Astronomy for Engineers..........................8vo, 2 0
Craig's Azimuth..4to, 3 0
Crandall's Text-book on Geodesy and Least Squares.................8vo, 3 50
Doolittle's Treatise on Practical Astronomy......................8vo, 4 00
Hayford's Text-book of Geodetic Astronomy........................8vo, 3 00
Hosmer's Azimuth...16mo, mor. 1 00
    Practical Astronomy. (In Press.)
Merriman's Elements of Precise Surveying and Geodesy.............8vo, 2 50
* Michie and Harlow's Practical Astronomy 8vo, 3 00
Rust's Ex-meridian Altitude, Azimuth and Star-Finding Tables.......8vo, 5 00
* White's Elements of Theoretical and Descriptive Astronomy 12mo, 2 00

CHEMISTRY.

* Abderhalden's Physiological Chemistry in Thirty Lectures. (Hall and
        Defren.)...8vo, 5 00
* Abegg's Theory of Electrolytic Dissociation. (von Ende.)........12mo, 1 25
Alexeyeff's General Principles of Organic Syntheses. (Matthews.)...8vo, 3 00
Allen's Tables for Iron Analysis.................................8vo, 3 00
Armsby's Principles of Animal Nutrition..........................8vo, 4 00
Arnold's Compendium of Chemistry. (Mandel.).............Large 12mo, 3 50

3

Association of State and National Food and Dairy **Departments, Hartford**
    Meeting, 1906...**8vo,**. $3 00
    Jamestown Meeting, 1907...........................,.....**8vo,** 3 00
Austen's Notes for Chemical Students....... ....**12mo,** 1 50
Baskerville's Chemical Elements. (In Preparation)   .
Bernadou's Smokeless Powder.—Nitro-cellulose, and Theory of the Cellulose
    Molecule...................................12mo, 2 50
Biltz's Introduction to Inorganic Chemistry. (Hall and Phelan.) ..12mo, 1 25
    Laboratory Methods of Inorganic Chemistry. (Hall and Blanchard.)
           8vo, 3 00
* Blanchard's Synthetic Inorganic Chemistry........................12mo, 1 00
* Browning's Introduction to the Rarer Elements................8vo, 1 50
* Claassen's Beet-sugar Manufacture. (Hall and Rolfe).... ....8vo, 3 00
Classen's Quantitative Chemical Analysis by Electrolysis. (Boltwood.).8vo, 3 00
Cohn's Indicators and Test-papers................................12mo, 2 00
    Tests and Reagents...................................8vo, 3 00
* Danneel's Electrochemistry. (Merriam.). ....,....................12mo, 1 25
Dannerth's Methods of Textile Chemistry.......................12mo, 2 00
Duhem's Thermodynamics and Chemistry. (Burgess.)...............8vo, 4 00
Effront's Enzymes and their Applications. (Prescott.)..............8vo, 3 00
Eissler's Modern High Explosives...................................8vo, 4 00
Erdmann's Introduction to Chemical Preparations. (Dunlap)......12mo, 1 25
Fischer's Oedema. (In Press.)
    * Physiology of Alimentation.. ....Large 12mo, 2 00
Fletcher's Practical Instructions in Quantitative Assaying with the Blowpipe.
           16mo, mor. 1 50
Fowler's Sewage Works Analyses.................................12mo, 2 00
Fresenius's Manual of Qualitative Chemical Analysis. (Wells.)........:.8vo, 5 00
    Manual of Qualitative Chemical Analysis. Part I. Descriptive. (Wells.)8vo, 3 00
    Quantitative Chemical Analysis. (Cohn.) 2 vols..............8vo, 12 50
        When Sold Separately, Vol. I, $6. Vol. II, $8.
Fuertes's Water and Public Health............................12mo, 1 50
Furman and Pardoe's Manual of Practical Assaying (Sixth Edition,
    Revised and Enlarged.)................................8vo, 3 00
* Getman's Exercises in Physical Chemistry.......................12mo, 2 00
Gill's Gas and Fuel Analysis for Engineers...... ....12mo, 1 25
* Gooch and Browning's Outlines of Qualitative Chemical Analysis.
           Large 12mo, 1 25
Grotenfelt's Principles of Modern Dairy Practice. (Woll.)..........12mo, 2 00
Groth's Introduction to Chemical Crystallography (Marshall). 12mo, 1 25
Hammarsten's Text-book of Physiological Chemistry. (Mandel.)......8vo, 4 00
Hanausek's Microscopy of Technical Products. (Winton.)..............8vo, 5 00
* Haskins and Macleod's Organic Chemistry......................12mo, 2 00
* Herrick's Denatured or Industrial Alcohol........................8vo, 4 00
Hinds's Inorganic Chemistry.....................................8vo, 3 00
    * Laboratory Manual for Students............................12mo, 1 00
* Holleman's Laboratory Manual of Organic Chemistry for Beginners.
    (Walker.)..12mo, 1 00
    Text-book of Inorganic Chemistry. (Cooper.)...................8vo, 2 50
    Text-book of Organic Chemistry. (Walker and Mott.)............8vo, 2 50
* Holley's Lead and Zinc Pigments.......................Large 12mo, 3 00
Holley and Ladd's Analysis of Mixed Paints, Color Pigments, and Varnishes.
           Large 12mo, 2 50
Hopkins's Oil-chemists' Handbook................................8vo, 3 00
Jackson's Directions for Laboratory Work in Physiological Chemistry..8vo, 1 25
Johnson's Rapid Methods for the Chemical Analysis of Special Steels, Steel-
    making Alloys and Graphite...........................Large 12mo, 3 00
Landauer's Spectrum Analysis. (Tingle.)...........................8vo, 3 00
Lassar-Cohn's Application of Some General Reactions to Investigations in
    Organic Chemistry. (Tingle.)...........................12mo, 1 00
Leach's Inspection and Analysis of Food with Special Reference to State
    Control...8vo, 7 50
Löb's Electrochemistry of Organic Compounds. (Lorenz.)............8vo, 3 00
Lodge's Notes on Assaying and Metallurgical Laboratory Experiments..8vo, 3 00
Low's Technical Method of Ore Analysis...........................8vo, 3 00
Lowe's Paint for Steel Structures.................................12mo, 1 00
Lunge's Techno-chemical Analysis. (Cohn.)......................12mo, 1 00

4

5

Turneaure and Russell's Public Water-supplies.........................8vo, $5 00
Van Deventer's Physical Chemistry for Beginners. (Boltwood.)......12mo, 1 50
Venable's Methods and Devices for Bacterial Treatment of Sewage......8vo, 3 00
Ward and Whipple's Freshwater Biology. (In Press.)
Ware's Beet-sugar Manufacture and Refining.   Vol. I.................8vo, 4 00
    "   "   "   "   "   Vol. II..............8vo, 5 00
Washington's Manual of the Chemical Analysis of Rocks...............8vo, 2 00
* Weaver's Military Explosives.......................................8vo, 3 00
Wells's Laboratory Guide in Qualitative Chemical Analysis...........8vo, 1 50
    Short Course in Inorganic Qualitative Chemical Analysis for Engineering
        Students...12mo, 1 50
    Text-book of Chemical Arithmetic.............................12mo, 1 25
Whipple's Microscopy of Drinking-water.............................8vo, 3 50
Wilson's Chlorination Process....................................12mo, 1 50
    Cyanide Processes...12mo, 1 50
Winton's Microscopy of Vegetable Foods............................8vo, 7 50
Zsigmondy's Colloids and the Ultramicroscope. (Alexander.)..Large 12mo, 3 00

CIVIL ENGINEERING.

BRIDGES AND ROOFS. HYDRAULICS MATERIALS OF ENGINEER-

ING. RAILWAY ENGINEERING.

Baker's Engineers' Surveying Instruments..........................12mo, 3 00
Bixby's Graphical Computing Table ....Paper 19½×24½ inches. . 25
Breed and Hosmer's Principles and Practice of Surveying Vol. I. Elemen-
    tary Surveying 8vo, 3 00
    Vol. II. Higher Surveying 8vo, 2 50
* Burr's Ancient and Modern Engineering and the Isthmian Canal......8vo, 3 50
Comstock's Field Astronomy for Engineers..........................8vo, 2 50
* Corthell's Allowable Pressure on Deep Foundations 12mo, 1 25
Crandall's Text-book on Geodesy and Least Squares.................8vo, 3 00
Davis's Elevation and Stadia Tables...............................8vo, 1 00
Elliott's Engineering for Land Drainage..........................12mo, 1 50
* Fiebeger's Treatise on Civil Engineering........................8vo, 5 00
Flemer's Photographic Methods and Instruments.....................8vo, 5 00
Folwell's Sewerage. (Designing and Maintenance.)..................8vo, 3 00
Freitag's Architectural Engineering 8vo, 3 50
French and Ives's Stereotomy......................................8vo, 2 50
Goodhue's Municipal Improvements.................................12mo, 1 50
* Hauch and Rice's Tables of Quantities for Preliminary Estimates...12mo, 1 25
Hayford's Text-book of Geodetic Astronomy.. ....8vo, 3 00
Hering's Ready Reference Tables (Conversion Factors.)... ......16mo, mor. 2 50
Hosmer's Azimuth....................................... 16mo, mor. 1 00
Howe's Retaining Walls for Earth.................................12mo, 1 25
* Ives's Adjustments of the Engineer's Transit and Level....... 16mo, bds. 25
Ives and Hilts's Problems in Surveying, Railroad Surveying and Geod-
    esy...16mo, mor. 1 50
Johnson's (J. B.) Theory and Practice of Surveying......... Large 12mo, 4 00
Johnson's (L. J.) Statics by Algebraic and Graphic Methods...........8vo, 2 00
Kinnicutt, Winslow and Pratt's Sewage Disposal. (In Press)
* Mahan's Descriptive Geometry....................................8vo, 1 50
Merriman's Elements of Precise Surveying and Geodesy..............8vo, 2 50
Merriman and Brooks's Handbook for Surveyors..............16mo, mor. 2 00
Nugent's Plane Surveying..8vo, 3
Ogden's Sewer Construction.......................................8vo, 3
    Sewer Design...12mo, 2
Parsons's Disposal of Municipal Refuse............................8vo, 2
Patton's Treatise on Civil Engineering.................8vo, half leather, 7
Reed's Topographical Drawing and Sketching.......................4to, 5
Rideal's Sewage and the Bacterial Purification of Sewage............8vo, 4
Riemer's Shaft-sinking under Difficult Conditions. (Corning and Peele.).8vo, 3
Siebert and Biggin's Modern Stone-cutting and Masonry..............8vo, 1 50
Smith's Manual of Topographical Drawing. (McMillan.)..............8vo, 2 50

Soper's Air and Ventilation of Subways..........................12mo, $2 50
* Tracy's Exercises in Surveying..........................12mo, mor. 1 00
Tracy's Plane Surveying...............................16mo, mor. 3 00
* Trautwine's Civil Engineer's Pocket-book....................16mo, mor. 5 00
Venable's Garbage Crematories in America..........................8vo, 2 00
   Methods and Devices for Bacterial Treatment of Sewage...........8vo, 3 00
Wait's Engineering and Architectural Jurisprudence...................8vo, 6 00
   Sheep, 6 50
   Law of Contracts.......................................8vo, 3 00
   Law of Operations Preliminary to Construction in Engineering and
    Architecture...8vo, 5 00
   Sheep, 5 50
Warren's Stereotomy—Problems in Stone-cutting.....................8vo, 2 50
* Waterbury's Vest-Pocket Hand-book of Mathematics for Engineers.
       $2\frac{7}{8} \times 5\frac{3}{8}$ inches. mor. 1 00
   * Enlarged Edition, Including Tables ....mor. 1 50
Webb's Problems in the Use and Adjustment of Engineering Instruments.
    16mo, mor. 1 25
Wilson's Topographic Surveying..................................8vo, 3 50

BRIDGES AND ROOFS.

Boller's Practical Treatise on the Construction of Iron Highway Bridges..8vo, 2 00
   * Thames River Bridge...............................Oblong paper, 5 00
Burr and Falk's Design and Construction of Metallic Bridges..........8vo, 5 00
   Influence Lines for Bridge and Roof Computations.................8vo, 3 00
Du Bois's Mechanics of Engineering. Vol. II...................Small 4to, 10 00
Foster's Treatise on Wooden Trestle Bridges........................4to, 5 00
Fowler's Ordinary Foundations......................................8vo, 3 50
Greene's Arches in Wood, Iron, and Stone..........................8vo, 2 50
   Bridge Trusses...8vo, 2 50
   Roof Trusses..8vo, 1 25
Grimm's Secondary Stresses in Bridge Trusses.........................8vo, 2 50
Heller's Stresses in Structures and the Accompanying Deformations....8vo, 3 00
Howe's Design of Simple Roof-trusses in Wood and Steel..............8vo. 2 00
   Symmetrical Masonry Arches....................................8vo, 2 50
   Treatise on Arches..8vo, 4 00
* Jacoby's Structural Details, or Elements of Design in Heavy Framing, 8vo, 2 25
Johnson, Bryan and Turneaure's Theory and Practice in the Designing of
   Modern Framed Structures..........................Small 4to, 10 00
* Johnson, Bryan and Turneaure's Theory and Practice in the Designing of
   Modern Framed Structures. New Edition. Part I. 8vo, 3 00
Merriman and Jacoby's Text-book on Roofs and Bridges:
   Part I.   Stresses in Simple Trusses...........................8vo, 2 50
   Part II. Graphic Statics.....................................8vo, 2 50
   Part III. Bridge Design......................................8vo, 2 50
   Part IV. Higher Structures....................................8vo, 2 50
Morison's Memphis Bridge....................................Oblong 4to, 10 00
Sondericker's Graphic Statics, with Applications to Trusses, Beams, and
   Arches...8vo, 2 00
Waddell's De Pontibus, Pocket-book for Bridge Engineers.......16mo, mor. 2 00
   * Specifications for Steel Bridges............................12mo, 50
Waddell and Harrington's Bridge Engineering. (In Preparation.)

HYDRAULICS.

Barnes's Ice Formation...8vo, 3 00
Bazin's Experiments upon the Contraction of the Liquid Vein Issuing from
   an Orifice. (Trautwine.)...................................8vo, 2 00
Bovey's Treatise on Hydraulics....................................8vo, 5 00
Church's Diagrams of Mean Velocity of Water in Open Channels.
    Oblong 4to, paper, 1 50
   Hydraulic Motors..8vo, 2 00

Coffin's Graphical Solution of Hydraulic Problems........... .16mo, mor. $2 50
Flather's Dynamometers, and the Measurement of Power........ ...12mo, 3 00
Folwell's Water-supply Engineering................................8vo, 4 00
Frizell's Water-power..................................... ...8vo, 5 00
Fuertes's Water and Public Health..............................12mo, 1 50
    Water-filtration Works 12mo, 2 50
Ganguillet and Kutter's General Formula for the Uniform Flow of Water in
    Rivers and Other Channels. (Hering and Trautwine.)......8vo, 4 00
Hazen's Clean Water and How to Get It...................Large 12mo, 1 50
    Filtration of Public Water-supplies.........................8vo, 3 00
Hazelhurst's Towers and Tanks for Water-works....................8vo, 2 50
Herschel's 115 Experiments on the Carrying Capacity of Large, Riveted, Metal
    Conduits..8vo, 2 00
Hoyt and Grover's River Discharge..............................8vo, 2 00
Hubbard and Kiersted's Water-works Management and Maintenance.
                                                            8vo, 4 00
* Lyndon's Development and Electrical Distribution of Water Power.
                                                            8vo, 3 00
Mason's Water-supply. (Considered Principally from a Sanitary Stand-
    point.)...8vo, 4 00
Merriman's Treatise on Hydraulics................................8vo, 5 00
* Molitor's Hydraulics of Rivers, Weirs and Sluices.................8vo, 2 00
* Morrison and Brodie's High Masonry Dam Design.................8vo, 1 50
* Richards's Laboratory Notes on Industrial Water Analysis.........8vo, 50
Schuyler's Reservoirs for Irrigation, Water-power, and Domestic Water-
    supply. Second Edition, Revised and Enlarged.......Large 8vo,— 6 00
* Thomas and Watt's Improvement of Rivers.......................4to, 6 00
Turneaure and Russell's Public Water-supplies....................8vo, 5 00
Wegmann's Design and Construction of Dams. 5th Ed., enlarged......4to, 6 00
    Water-Supply of the City of New York from 1658 to 1895.........4to, 10 00
Whipple's Value of Pure Water....................... ......Large 12mo, 1 00
Williams and Hazen's Hydraulic Tables...........................8vo, 1 50
Wilson's Irrigation Engineering..................................8vo, 4 00
Wood's Turbines..8vo, 2 50

MATERIALS OF ENGINEERING.

Baker's Roads and Pavements.....................................8vo, 5 00
    Treatise on Masonry Construction.............................8vo, 5 00
Black's United States Public Works.........................Oblong 4to, 5 00
Blanchard's Bituminous Roads. (In Preparation.)
Bleininger's Manufacture of Hydraulic Cement. (In Preparation.)
* Bovey's Strength of Materials and Theory of Structures...............8vo, 7 50
Burr's Elasticity and Resistance of the Materials of Engineering.. .....8vo, 7 50
Byrne's Highway Construction.....................................8vo, 5 00
    Inspection of the Materials and Workmanship Employed in Construction.
                                                            16mo, 3 00
Church's Mechanics of Engineering................................8vo, 6 00
Du Bois's Mechanics of Engineering.
    Vol. I Kinematics, Statics, Kinetics....................Small 4to, 7 50
    Vol. II. The Stresses in Framed Structures, Strength of Materials and
        Theory of Flexures...............................Small 4to, 10 00
* Eckel's Cements, Limes, and Plasters...........................8vo, 6 00
    Stone and Clay Products used in Engineering. (In Preparation.)
Fowler's Ordinary Foundations...................................8vo, 3 50
* Greene's Structural Mechanics..................................8vo, 2 50
* Holley's Lead and Zinc Pigments......................Large 12mo, 3 00
Holley and Ladd's Analysis of Mixed Paints, Color Pigments and Varnishes.
                                                        Large 12mo, 2 50
* Hubbard's Dust Preventives and Road Binders..................8vo, 3 00
Johnson's (C. M.) Rapid Methods for the Chemical Analysis of Special Steels,
        Steel-making Alloys and Graphite.................Large 12mo, 3 00
Johnson's (J. B.) Materials of Construction....................Large 8vo, 6 00
Keep's Cast Iron..8vo, 2 50
Lanza's Applied Mechanics..8vo, 7 50
Lowe's Paints for Steel Structures..............................12mo, 1 00

8

Maire's Modern Pigments and their Vehicles........................12mo, $2 00
Maurer's Technical Mechanics...8vo, 4 00
Merrill's Stones for Building and Decoration.......................8vo, 5 00
Merriman's Mechanics of Materials.................................8vo, 5 00
   * Strength of Materials..12mo, 1 00
Metcalf's Steel. A Manual for Steel-users........................12mo, 2 00
Morrison's Highway Engineering......................................8vo, 2 50
Patton's Practical Treatise on Foundations.........................8vo, 5 00
Rice's Concrete Block Manufacture..................................8vo, 2 00
Richardson's Modern Asphalt Pavement..............................8vo, 3 00
Richey's Building Foreman's Pocket Book and Ready Reference.16mo,mor. 5 00
   * Cement Workers' and Plasterers' Edition (Building Mechanics' Ready
      Reference Series)...16mo, mor. 1 50
   Handbook for Superintendents of Construction.............16mo, mor. 4 00
   * Stone and Brick Masons' Edition (Building Mechanics' Ready
      Reference Series)...16mo, mor. 1 50
* Ries's Clays: Their Occurrence, Properties, and Uses.................8vo, 5 00
* Ries and Leighton's History of the Clay-working Industry of the United
      States...8vo, 2 50
Sabin's Industrial and Artistic Technology of Paint and Varnish........8vo, 3 00
* Smith's Strength of Material....................................12mo 1 25
Snow's Principal Species of Wood..................................8vo, 3 50
Spalding's Hydraulic Cement......................................12mo, 2 00
   Text-book on Roads and Pavements.............................12mo, 2 00
*'Taylor and Thompson's Extracts on Reinforced Concrete Designs....8vo, 2 50
   Treatise on Concrete, Plain and Reinforced.. ....8vo, 5 00
Thurston's Materials of Engineering. In Three Parts.................8vo, 8 00
   Part I. Non-metallic Materials of Engineering and Metallurgy....8vo, 2 00
   Part II. Iron and Steel...8vo, 3 50
   Part III. A Treatise on Brasses, Bronzes, and Other Alloys and their
      Constituents...8vo, 2 50
Tillson's Street Pavements and Paving Materials....................8vo, 4 00
* Trautwine's Concrete, Plain and Reinforced 16mo, 2 00
Turneaure and Maurer's Principles of Reinforced Concrete Construction.
   Second Edition, Revised and Enlarged....................8vo, 3 50
Waterbury's Cement Laboratory Manual..........................12mo, 1 00
Wood's (De V.) Treatise on the Resistance of Materials, and an Appendix on
      the Preservation of Timber.................................8vo, 2 00
Wood's (M. P.) Rustless Coatings: Corrosion and Electrolysis of Iron and
      Steel..8vo, 4 00

RAILWAY ENGINEERING.

Andrews's Handbook for Street Railway Engineers.......3×5 inches, mor 1 25
Berg's Buildings and Structures of American Railroads..............4to, 5 00
Brooks's Handbook of Street Railroad Location.............16mo, mor. 1 50
Butts's Civil Engineer's Field-book............................16mc, mor. 2 50
Crandall's Railway and Other Earthwork Tables.....................8vo, 1 50
   Transition Curve...16mo, mor. 1 50
* Crockett's Methods for Earthwork Computations...................8vo, 1 50
Dredge's History of the Pennsylvania Railroad. (1879)...............Paper, 5 00
Fisher's Table of Cubic Yards.................................Cardboard, 25
Godwin's Railroad Engineers' Field-book and Explorers' Guide..16mo, mor. 2 50
Hudson's Tables for Calculating the Cubic Contents of Excavations and Em-
   bankments..... ....8vo, 1 00
Ives and Hilts's Problems in Surveying, Railroad Surveying and Geodesy
      16mo, mor. 1 50
Molitor and Beard's Manual for Resident Engineers........... .....16mo, 1 00
Nagle's Field Manual for Railroad Engineers..................16mo, mor. 3 00
* Orrock's Railroad Structures and Estimates...................8vo, 3 00
Philbrick's Field Manual for Engineers.......................16mo, mor. 3 00
Raymond's Railroad Engineering. 3 volumes.
   Vol. I. Railroad Field Geometry. (In Press.)
   Vol. II. Elements of Railroad Engineering....................8vo, 3 50
   Vol. III. Railroad Engineer's Field Book. (In Preparation)
9

Roberts' Track Formulæ and Tables. (In Press)
Searles's Field Engineering..................................16mo, mor. $3 00
    Railroad Spiral...................................16mo, mor. 1 50
Taylor's Prismoidal Formulæ and EarthWork.. ....8vo, 1 50
* Trautwine's Field Practice of Laying Out Circular Curves for Railroads.
                                                  12mo, mor. 2 50
    * Method of Calculating the Cubic Contents of Excavations and Em-
      bankments by the Aid of Diagrams 8vo, 2 00
Webb's Economics of Railroad Construction...................Large 12mo, 2 50
    Railroad Construction.......................................16mo, mor. 5 00
Wellington's Economic Theory of the Location of Railways.....Large 12mo, 5 00
Wilson's Elements of Railroad-Track and Construction..............12mo, 2 00

DRAWING.

Barr's Kinematics of Machinery......................................8vo, 2 50
* Bartlett's Mechanical Drawing 8vo, 3 00
*   "   "   "   Abridged Ed......................8vo, 1 50
Bartlett and Johnson's Engineering Descriptive Geometry. (In Press)
Coolidge's Manual of Drawing...............................8vo, paper, 1 00
Coolidge and Freeman's Elements of General Drafting for Mechanical Engi-
    neers...Oblong 4to, 2 50
Durley's Kinematics of Machines.................................8vo, 4 00
Emch's Introduction to Projective Geometry and its Application......8vo, 2 50
Hill's Text-book on Shades and Shadows, and Perspective 8vo, 2 00
Jamison's Advanced Mechanical Drawing........................8vo, 2 00
    Elements of Mechanical Drawing...........................8vo, 2 50
Jones's Machine Design:
    Part I. Kinematics of Machinery............................8vo, 1 50
    Part II. Form, Strength, and Proportions of Parts................8vo, 3 00
* Kimball and Barr's Machine Design 8vo, 3 00
MacCord's Elements of Descriptive Geometry......................8vo, 3 00
    Kinematics; or, Practical Mechanism.........................8vo, 5 00
    Mechanical Drawing..4to, 4 00
    Velocity Diagrams...8vo, 1 50
McLeod's Descriptive Geometry...........................Large 12mo, 1 50
* Mahan's Descriptive Geometry and Stone-cutting.................8vo, 1 50
    Industrial Drawing. (Thompson.)...........................8vo, 3 50
Moyer's Descriptive Geometry.....................................8vo, 2 00
Reed's Topographical Drawing and Sketching......................4to, 5 00
Reid's Course in Mechanical Drawing............................8vo, 2 00
    Text-book of Mechanical Drawing and Elementary Machine Design .8vo, 3 00
Robinson's Principles of Mechanism.............................8vo, 3 00
Schwamb and Merrill's Elements of Mechanism....................8vo, 3 00
Smith (A. W.) and Marx's Machine Design.......................8vo, 3 00
Smith's (R. S.) Manual of Topographical Drawing. (McMillan)........8vo, 2 50
* Titsworth's Elements of Mechanical Drawing...............Oblong 8vo, 1 25
Warren's Drafting Instruments and Operations...................12mo, 1 25
    Elements of Descriptive Geometry, Shadows, and Perspective.:....8vo, 3 50
    Elements of Machine Construction and Drawing.................8vo, 7 50
    Elements of Plane and Solid Free-hand Geometrical Drawing....12mo, 1 00
    General Problems of Shades and Shadows.....................8vo, 3 00
    Manual of Elementary Problems in the Linear Perspective of Forms and
      Shadow...12mo, 1 00
    Manual of Elementary Projection Drawing....................12mo, 1 50
    Plane Problems in Elementary Geometry......................12mo, 1 25
Weisbach's Kinematics and Power of Transmission. (Hermann and
    Klein.)...8vo, 5 00
Wilson's (H. M.) Topographic Surveying.........................8vo, 3 50
* Wilson's (V. T.) Descriptive Geometry.........................8vo, 1 50
    Free-hand Lettering.......................................8vo, 1 00
    Free-hand Perspective.....................................8vo, 2 50
Woolf's Elementary Course in Descriptive Geometry...........Large 8vo, 3 00

10

ELECTRICITY AND PHYSICS.

* Abegg's Theory of Electrolytic Dissociation. (von Ende.).........12mo, $1 25
Andrews's Hand-book for Street Railway Engineering.....3×5 inches, mor. 1 25
Anthony and Ball's Lecture-notes on the Theory of Electrical Measure-
    ments 12mo, 1 00
Anthony and Brackett's Text-book of Physics. (Magie.)....Large 12mo, 3 00
Benjamin's History of Electricity..................................8vo, 3 00
Betts's Lead Refining and Electrolysis.............................8vo, 4 00
Classen's Quantitative Chemical Analysis by Electrolysis. (Boltwood.).8vo, 3 00
* Collins's Manual of Wireless Telegraphy and Telephony............12mo, 1 50
Crehore and Squier's Polarizing Photo-chronograph..................8vo, 3 00
* Danneel's Electrochemistry. (Merriam.).........................12mo, 1 25
Dawson's "Engineering" and Electric Traction Pocket-book....16mo, mor. 5 00
Dolezalek's Theory of the Lead Accumulator (Storage Battery). (von Ende.)
    12mo, 2 50
Duhem's Thermodynamics and Chemistry. (Burgess.)................8vo, 4 00
Flather's Dynamometers, and the Measurement of Power............12mo, 3 00
* Getman's Introduction to Physical Science. 12mo, 1 50
Gilbert's De Magnete. (Mottelay.).................................8vo, 2 50
* Hanchett's·Alternating Currents.................................12mo, 1 00
Hering's Ready Reference Tables (Conversion Factors)........16mo, mor. 2 50
* Hobart and Ellis's High-speed Dynamo Electric Machinery..........8vo, 6 00
Holman's Precision of Measurements................................8vo, 2 -00
    Telescopic Mirror-scale Method, Adjustments, and Tests....Large 8vo, 75
* Karapetoff's Experimental Electrical Engineering..................8vo, 6 00
Kinzbrunner's Testing of Continuous-current Machines..............8vo, 2 00
Landauer's Spectrum Analysis. (Tingle.).........................8vo, 3 00
Le Chatelier's High-temperature Measurements. (Boudouard—Burgess)12mo, 3 00
Lob's Electrochemistry of Organic Compounds. (Lorenz.) 8vo, 3 00
* Lyndon's Development and Electrical Distribution of Water Power..8vo, 3 00
* Lyons's Treatise on Electromagnetic Phenomena. Vols, I .and II. 8vo, each, 6 00
* Michie's Elements of Wave Motion Relating to Sound and Light.....8vo, 4 00
Morgan's Outline of the Theory of Solution and its Results..........12mo, 1 00
    * Physical Chemistry for Electrical Engineers..................12mo, 1 50
* Norris's Introduction to the Study of Electrical Engineering........8vo, 2 50
Norris and Dennison's Course of Problems on the Electrical Characteristics of
    Circuits and Machines. (In Press.)
* Parshall and Hobart's Electric Machine Design..........4to, half mor, 12 50
Reagan's Locomotives: Simple, Compound, and Electric. New Edition
    Large 12mo, 3 50
* Rosenberg's Electrical Engineering. (Haldane Gee—Kinzbrunner.)..8vo, 2 00
Ryan, Norris, and Hoxie's Electrical Machinery. Vol. I..............8vo, 2 50
Schapper's Laboratory Guide for Students in Physical Chemistry.....12mo, 1 00
* Tillman's Elementary Lessons in Heat............................8vo, 1 50
Tory and Pitcher's Manual of Laboratory Physics............Large 12mo, 2 00
Ulke's Modern Electrolytic Copper Refining........................8vo, 3 00

LAW.

* Brennan's Hand-book of Useful Legal Information for Business Men.
    16mo, mor. 5 00
* Davis's Elements of Law..8vo, 2 50
    * Treatise on the Military Law of United States................8vo, 7 00
* Dudley's Military Law and the Procedure of Courts-martial..Large 12mo, 2 50
Manual for Courts-martial....................................16mo, mor. 1 50
Wait's Engineering and Architectural Jurisprudence..................8vo, 6 00
    Sheep, 6 50
    Law of Contracts...8vo, 3 00
    Law of Operations Preliminary to Construction in Engineering and
        Architecture...8vo, 5 00
    Sheep, 5 50
Baker's Elliptic Functions...8vo, 1 50

11

MATHEMATICS.

Briggs's Elements of Plane Analytic Geometry. (Bôcher.) 12mo, $1 00
* Buchanan's Plane and Spherical Trigonometry.....................8vo, 1 00
Byerly's Harmonic Functions..8vo, 1 00
Chandler's Elements of the Infinitesimal Calculus..................12mo, 2 00
* Coffin's Vector Analysis. 12mo, 2 50
Compton's Manual of Logarithmic Computations...................12mo, 1 50
* Dickson's College Algebra..............................Large 12mo, 1 50
        * Introduction to the Theory of Algebraic Equations......Large 12mo, 1 25
Emch's Introduction to Projective Geometry and its Application......8vo, 2 50
Fiske's Functions of a Complex Variable...........................8vo, 1 00
Halsted's Elementary Synthetic Geometry.........................8vo, 1 50
        Elements of Geometry...8vo, 1 75
        * Rational Geometry..12mo, 1 50
        Synthetic Projective Geometry................................8vo, 1 00
* Hancock's Lectures on the Theory of Elliptic Functions 8vo, 5 00
Hyde's Grassmann's Space Analysis...............................8vo, 1 00
* Johnson's (J. B.) Three-place Logarithmic Tables: Vest-pocket size, paper, 15
                                    * 100 copies, 5 00
                * Mounted on heavy cardboard, 8 × 10 inches, 25
                                    * 10 copies, 2 00
Johnson's (W. W.) Abridged Editions of Differential and Integral Calculus.
                                    Large 12mo, 1 vol. 2 50
        Curve Tracing in Cartesian Co-ordinates.....................12mo, 1
        Differential Equations...8vo, 1
        Elementary Treatise on Differential Calculus..............Large 12mo, 1
        Elementary Treatise on the Integral Calculus...........Large 12mo, 1
        * Theoretical Mechanics.......................................12mo, 3
        Theory of Errors and the Method of Least Squares............12mo, 1
        Treatise on Differential Calculus.......................Large 12mo, 3
        Treatise on the Integral Calculus.......................Large 12mo, 3 00
        Treatise on Ordinary and Partial Differential Equations...Large 12mo, 3 50
Karapetoff's Engineering Applications of Higher Mathematics. (In Preparation.)
Laplace's Philosophical Essay on Probabilities. (Truscott and Emory.). 12mo, 2 00
* Ludlow's Logarithmic and Trigonometric Tables..................8vo, 1 00
* Ludlow and Bass's Elements of Trigonometry and Logarithmic and Other
        Tables. 8vo, 3
        * Trigonometry and Tables published separately. ....Each, 2
Macfarlane's Vector Analysis and Quaternions.....................8vo, 1 00
McMahon's Hyperbolic Functions. 8vo, 1 00
Manning's Irrational Numbers and their Representation by Sequences and
        Series. ,...12mo, 1 25
Mathematical Monographs. Edited by Mansfield Merriman and Robert
        S. Woodward.................................Octavo, each 1 00
        No. 1. History of Modern Mathematics, by David Eugene Smith.
        No. 2. Synthetic Projective Geometry, by George Bruce Halsted.
        No. 3. Determinants, by Laenas Gifford Weld. No. 4. Hyper-
        bolic Functions, by James McMahon. No. 5. Harmonic Func-
        tions, by William E. Byerly. No. 6. Grassmann's Space Analysis,
        by Edward W. Hyde. No. 7. Probability and Theory of Errors,
        by Robert S. Woodward. No. 8. Vector Analysis and Quaternions,
        by Alexander Macfarlane. No. 9. Differential Equations, by
        William Woolsey Johnson. No. 10. The Solution of Equations,
        by Mansfield Merriman. No. 11. Functions of a Complex Variable,
        by Thomas S. Fiske.
Maurer's Technical Mechanics......................................8vo, 4 00
Merriman's Method of Least Squares. 8vo, 2 00
        Solution of Equations...8vo, 1 00
Moritz's Elements of Plane Trigonometry. (In Press.)
Rice and Johnson's Differential and Integral Calculus. 2 vols. in one.
                                    Large 12mo, 1 50
        Elementary Treatise on the Differential Calculus. ....Large 12mo, 3 00
Smith's History of Modern Mathematics.............................8vo, 1 00
* Veblen and Lennes's Introduction to the Real Infinitesimal Analysis of One
        Variable...8vo, 2 00

* Waterbury's Vest Pocket Hand-book of Mathematics for Engineers

$2\frac{1}{8} \times 5\frac{3}{8}$ inches, mor. $1 00

    * Enlarged Edition, Including Tables ....mor. 1 50

Weld's Determinants...8vo, 1 00

Wood's Elements of Co-ordinate Geometry........................8vo, 2 00

Woodward's Probability and Theory of Errors....................8vo, 1 00

MECHANICAL ENGINEERING.

MATERIALS OF ENGINEERING, STEAM-ENGINES AND BOILERS.

Bacon's Forge Practice..12mo, 1 50

Baldwin's Steam Heating for Buildings...........................12mo, 2 50

Barr's Kinematics of Machinery..................................8vo, 2 50

* Bartlett's Mechanical Drawing..................................8vo, 3 00

*   "   "   " Abridged Ed......................8vo, 1 50

Bartlett and Johnson's Engineering Descriptive Geometry. (In Press.)

* Burr's Ancient and Modern Engineering and the Isthmian Canal.....8vo, 3 50

Carpenter's Experimental Engineering............................8vo, 6 00

    Heating and Ventilating Buildings...........................8vo, 4 00

* Clerk's The Gas, Petrol and Oil Engine 8vo, 4 00

Compton's First Lessons in Metal Working........................12mo, 1 50

Compton and De Groodt's Speed Lathe............................12mo, 1 50

Coolidge's Manual of Drawing.............................. 8vo, paper, 1 00

Coolidge and Freeman's Elements of General Drafting for Mechanical Engineers...Oblong 4to, 2 50

Cromwell's Treatise on Belts and Pulleys........................12mo, 1 50

    Treatise on Toothed Gearing.................................12mo, 1 50

Dingey's Machinery Pattern Making...............................12mo, 2 00

Durley's Kinematics of Machines.................................8vo, 4 00

Flanders's Gear-cutting Machinery. ....Large 12mo, 3 00

Flather's Dynamometers and the Measurement of Power........12mo, 3 00

    Rope Driving..12mo, 2 00

Gill's Gas and Fuel Analysis for Engineers......................12mo, 1 25

Goss's Locomotive Sparks..8vo, 2 00

Greene's Pumping Machinery. (In Preparation.)

Hering's Ready Reference Tables (Conversion Factors)........16mo, mor. 2 50

* Hobart and Ellis's High Speed Dynamo Electric Machinery........8vo, 6 00

Hutton's Gas Engine...8vo, 5 00

Jamison's Advanced Mechanical Drawing...........................8vo, 2 00

    Elements of Mechanical Drawing..............................8vo, 2 50

Jones's Gas Engine..8vo, 4 00

    Machine Design:

    Part I.   Kinematics of Machinery...........................8vo, 1 50

    Part II.   Form, Strength, and Proportions of Parts...........8vo, 3 00

* Kent's Mechanical Engineer's Pocket-Book.................16mo, mor. 5 00

Kerr's Power and Power Transmission.............................8vo, 2 00

* Kimball and Barr's Machine Design..............................8vo, 3 00

Leonard's Machine Shop Tools and Methods........................8vo, 4 00

* Levin's Gas Engine...8vo, 4 00

* Lorenz's Modern Refrigerating Machinery. (Pope, Haven, and Dean)..8vo, 4 00

MacCord's Kinematics; or, Practical Mechanism...................8vo, 5 00

    Mechanical Drawing..4to, 4 00

    Velocity Diagrams...8vo, 1 50

MacFarland's Standard Reduction Factors for Gases..............8vo, 1 50

Mahan's Industrial Drawing. (Thompson.)........................8vo, 3 50

Mehrtens's Gas Engine Theory and Design..................Large 12mo, 2 50

Oberg's Handbook of Small Tools.........................Large 12mo, 3 00

* Parshall and Hobart's Electric Machine Design. Small 4to, half leather, 12 50

Peele's Compressed Air Plant for Mines.........................8vo, 3 00

Poole's Calorific Power of Fuels................................8vo, 3 00

* Porter's Engineering Reminiscences, 1855 to 1882..............8vo, 3 00

Reid's Course in Mechanical Drawing............................8vo, 2 00

    Text-book of Mechanical Drawing and Elementary Machine Design.8vo, 3 00

Richards's Compressed Air..12mo, $1 50
Robinson's Principles of Mechanism...............................8vo, 00
Schwamb and Merrill's Elements of Mechanism...................8vo, 00
Smith (A. W.) and Marx's Machine Design.......................8vo, 00
Smith's (O.) Press-working of Metals.............................8vo, 3 00
Sorel's Carbureting and Combustion in Alcohol Engines. (Woodward and
    Preston.)....................................Large 12mo, 00
Stone's Practical Testing of Gas and Gas Meters...................8vo, 3 50
Thurston's Animal as a Machine and Prime Motor, and the Laws of Energetics.
    12mo, 1 00
    Treatise on Friction and Lost Work in Machinery and Mill Work...8vo, 3 00
* Tillson's Complete Automobile Instructor.......................16mo, 1 50
* Titsworth's Elements of Mechanical Drawing.............Oblong 8vo, 1 25
Warren's Elements of Machine Construction and Drawing...........8vo, 7 50
* Waterbury's Vest Pocket Hand-book of Mathematics for Engineers.
    $2\frac{7}{8} \times 5\frac{3}{8}$ inches, mor. 1 00
    * Enlarged Edition, Including Tables........................mor. 1 50
Weisbach's Kinematics and the Power of Transmission. (Herrmann—
    Klein)...8vo, 5 00
    Machinery of Transmission and Governors. (Hermann—Klein.)..8vo, 5 00
Wood's Turbines...8vo, 2 50

MATERIALS OF ENGINEERING.

* Bovey's Strength of Materials and Theory of Structures...........8vo, 7 50
Burr's Elasticity and Resistance of the Materials of Engineering.......8vo, 7 50
Church's Mechanics of Engineering...............................8vo, 6 00
* Greene's Structural Mechanics.................................8vo, 2 50
* Holley's Lead and Zinc Pigments........................Large 12mo 3 00
Holley and Ladd's Analysis of Mixed Paints, Color Pigments, and Varnishes.
    Large 12mo, 2 50
Johnson's (C. M.) Rapid Methods for the Chemical Analysis of Special
    Steels, Steel-Making Alloys and Graphite...........Large 12mo, 3 00
Johnson's (J. B.) Materials of Construction......................8vo, 6 00
Keep's Cast Iron..8vo, 2 50
Lanza's Applied Mechanics......................................8vo, 7 50
Lowe's Paints for Steel Structures..............................12mo, 1 00
Maire's Modern Pigments and their Vehicles......................12mo, 2 00
Maurer's Technical Mechanics...................................8vo, 4 00
Merriman's Mechanics of Materials..............................8vo, 5 00
    * Strength of Materials.....................................12mo, 1 00
Metcalf's Steel. A Manual for Steel-users........................12mo, 2 00
Sabin's Industrial and Artistic Technology of Paint and Varnish......8vo, 3 00
Smith's ((A. W.) Materials of Machines...........................12mo, 1 00
* Smith's (H. E.) Strength of Material............................12mo, 1 25
Thurston's Materials of Engineering.......................3 vols., 8vo, 8 00
    Part I. Non-metallic Materials of Engineering,.................8vo, 2 00
    Part II. Iron and Steel......................................8vo, 3 50
    Part III. A Treatise on Brasses, Bronzes, and Other Alloys and their
    Constituents...8vo, 2 50
Wood's (De V.) Elements of Analytical Mechanics.................8vo, 3 00
    Treatise on the Resistance of Materials and an Appendix on the
    Preservation of Timber....................................8vo, 2 00
Wood's (M. P.) Rustless Coatings: Corrosion and Electrolysis of Iron and
    Steel...8vo, 4 00

STEAM-ENGINES AND BOILERS.

Berry's Temperature-entropy Diagram............................12mo, 2 00
Carnot's Reflections on the Motive Power of Heat. (Thurston.)......12mo, 1 50
Chase's Art of Pattern Making..................................12mo, 2 50

Creighton's Steam-engine and other Heat Motors....................8vo, $5 00
Dawson's "Engineering" and Electric Traction Pocket-book. ....16mo, mor.   00
* Gebhardt's Steam Power Plant Engineering.......................8vo,   00
Goss's Locomotive Performance....................................8vo,   00
Hemenway's Indicator Practice and Steam-engine Economy.........12mo,   00
Hutton's Heat and Heat-engines..................................8vo,   00
    Mechanical Engineering of Power Plants......................8vo, 5 00
Kent's Steam Boiler Economy 8vo, 4 00
Kneass's Practice and Theory of the Injector....................8vo, 1 50
MacCord's Slide-valves..8vo, 2 00
Meyer's Modern Locomotive Construction..........................4to, 10 00
Moyer's Steam Turbine..8vo, 4 00
Peabody's Manual of the Steam-engine Indicator..................12mo, 1 50
    Tables of the Properties of Steam and Other Vapors and Temperature-
      Entropy Table...8vo, 1 00
    Thermodynamics of the Steam-engine and Other Heat-engines. ...8vo,   00
    Valve-gears for Steam-engines...............................8vo,   50
Peabody and Miller's Steam-boilers..............................8vo, 4 00
Pupin's Thermodynamics of Reversible Cycles in Gases and Saturated Vapors.
    (Osterberg.)..12mo, 1 25
Reagan's Locomotives: Simple, Compound, and Electric.   New Edition.
                                          Large 12mo, 3 50
Sinclair's Locomotive Engine Running and Management...........12mo, 2 00
Smart's Handbook of Engineering Laboratory Practice............12mo, 2 50
Snow's Steam-boiler Practice....................................8vo, 3 00
Spangler's Notes on Thermodynamics...........................12mo, 1 00
    Valve-gears...8vo, 2 50
Spangler, Greene, and Marshall's Elements of Steam-engineering......8vo, 3 00
Thomas's Steam-turbines...8vo, 4 00
Thurston's Handbook of Engine and Boiler Trials, and the Use of the Indi-
    cator and the Prony Brake...............................8vo, 5 00
    Handy Tables..8vo, 1 50
    Manual of Steam-boilers, their Designs, Construction, and Operation 8vo, 5 00
    Manual of the Steam-engine..........................2 vols., 8vo, 10 00
        Part I.   History, Structure, and Theory 8vo, 6 00
        Part II.   Design, Construction, and Operation..............8vo, 6 00
Wehrenfennig's Analysis and Softening of Boiler Feed-water.   (Patterson.)
                                               8vo, 4 00
Weisbach's Heat, Steam, and Steam-engines.   (Du Bois.).......8vo, 5 00
Whitham's Steam-engine Design..................................8vo, 5 00
Wood's Thermodynamics, Heat Motors, and Refrigerating Machines...8vo, 4 00

MECHANICS PURE AND APPLIED.

Church's Mechanics of Engineering................................8vo, 6 00
    Notes and Examples in Mechanics............................8vo, 2 00
Dana's Text-book of Elementary Mechanics for Colleges and Schools .12mo, 1 50
Du Bois's Elementary Principles of Mechanics:
    Vol. I.   Kinematics......................................8vo, 3 50
    Vol. II.   Statics..8vo, 4 00
    Mechanics of Engineering.   Vol. I.......................Small 4to, 7 50
                 Vol. II.....................Small 4to, 10 00
* Greene's Structural Mechanics..................................8vo, 2 50
Hortmann's Elementary Mechanics for Engineering Students.   (In Press.)
James's Kinematics of a Point and the Rational Mechanics of a Particle.
                                        Large 12mo, 2 00
* Johnson's (W. W.) Theoretical Mechanics......................12mo, 3 00
Lanza's Applied Mechanics.......................................8vo, 7 50
* Martin's Text Book on Mechanics, Vol. I, Statics...............12mo, 1 25
                * Vol. II, Kinematics and Kinetics.12mo, 1 50
Maurer's Technical Mechanics....................................8vo, 4 00
* Merriman's Elements of Mechanics............................12mo, 1 00
    Mechanics of Materials.....................................8vo, 5 00
* Michie's Elements of Analytical Mechanics......................8vo, 4 00
Robinson's Principles of Mechanism..............................8vo, 3 00

15

Sanborn's Mechanics Problems. . . . . .Large 12mo, $1 50
Schwamb and Merrill's Elements of Mechanism. 8vo, 3 00
Wood's Elements of Analytical Mechanics. 8vo, 3 00
   Principles of Elementary Mechanics. .12mo, 1 25

MEDICAL.

* Abderhalden's Physiological Chemistry in Thirty Lectures. (Hall and
   Defren.). 8vo, 5
von Behring's Suppression of Tuberculosis. (Bolduan.). 12mo, 1
Bolduan's Immune Sera. 12mo, 1
Bordet's Studies in Immunity. (Gay) 8vo, 6 00
Chapin's The Sources and Modes of Infection. (In Press.)
Davenport's Statistical Methods with Special Reference to Biological Varia-
   tions. . . . . .16mo, mor. 1 50
Ehrlich's Collected Studies on Immunity. (Bolduan.). 8vo, 6 00
Fischer's Oedema. (In Press.)
   * Physiology of Alimentation. . . . . .Large 12mo, 2 00
de Fursac's Manual of Psychiatry. (Rosanoff and Collins.). . . .Large 12mo, 2 50
Hammarsten's Text-book on Physiological Chemistry. (Mandel.). 8vo, 4 00
Jackson's Directions for Laboratory Work in Physiological Chemistry. .8vo, 1 25
Lassar-Cohn's Praxis of Urinary Analysis. (Lorenz.). 12mo, 1 00
Mandel's Hand-book for the Bio-Chemical Laboratory. 12mo, 1 50
* Nelson's Analysis of Drugs and Medicines. 12mo, 3 00
* Pauli's Physical Chemistry in the Service of Medicine. (Fischer.). .12mo, 1 25
* Pozzi-Escot's Toxins and Venoms and their Antibodies. (Cohn.). .12mo, 1 00
Rostoski's Serum Diagnosis. (Bolduan.). 12mo, 1 00
Ruddiman's Incompatibilities in Prescriptions. 8vo, 2 00
   Whys in Pharmacy. 12mo, 1 00
Salkowski's Physiological and Pathological Chemistry. (Orndorff.) 8vo, 2 50
* Satterlee's Outlines of Human Embryology. 12mo, 1 25
Smith's Lecture Notes on Chemistry for Dental Students. 8vo, 2 50
* Whipple's Tyhpoid Fever. . . . . .Large 12mo, 3 00
* Woodhull's Military Hygiene for Officers of the Line . . . . .Large 12mo, 1 50
   * Personal Hygiene. 12mo, 1 00
Worcester and Atkinson's Small Hospitals Establishment and Maintenance,
   and Suggestions for Hospital Architecture, with Plans for a Small
   Hospital. 12mo, 1 25

METALLURGY.

Betts's Lead Refining by Electrolysis. 8vo, 4 00
Bolland's Encyclopedia of Founding and Dictionary of Foundry Terms used
   in the Practice of Moulding. 12mo, 3 00
   Iron Founder. 12mo, 2 50
   " " Supplement. 12mo, 2 50
Douglas's Untechnical Addresses on Technical Subjects. 12mo, 1 00
Goesel's Minerals and Metals: A Reference Book. . . . . .16mo, mor. 3 00
* Iles's Lead-smelting. 12mo, 2 50
Johnson's Rapid Methods for the Chemical Analysis of Special Steels,
   Steel-making Alloys and Graphite. . . . . .Large 12mo, 3 00
Keep's Cast Iron. 8vo, 2 50
Le Chatelier's High-temperature Measurements. (Boudouard—Burgess.)
   12mo, 3 00
Metcalf's Steel. A Manual for Steel-users. 12mo, 2 00
Minet's Production of Aluminum and its Industrial Use. (Waldo.). . 12mo, 2 50
* Ruer's Elements of Metallography. (Mathewson.). 8vo, 3 00
Smith's Materials of Machines. 12mo, 1 00
Tate and Stone's Foundry Practice. 12mo, 2 00
Thurston's Materials of Engineering. In Three Parts. 8vo, 8 00
   Part I. Non-metallic Materials of Engineering, see Civil Engineering,
   page 9.
   Part II. Iron and Steel. 8vo, 3 50
   Part III. A Treatise on Brasses, Bronzes, and Other Alloys and their
   Constituents. 8vo, 2 50

Uike's Modern Electrolytic Copper Refining. 8vo, $3 00
West's American Foundry Practice. 12mo, 1 50
    Moulders' Text Book. 12mo, 2 50

MINERALOGY.

Baskerville's Chemical Elements. (In Preparation.)
* Browning's Introduction to the Rarer Elements. 8vo, 1 50
Brush's Manual of Determinative Mineralogy. (Penfield.). 8vo, 4 00
Butler's Pocket Hand-book of Minerals. . . . . : . . . . .16.no, mor. 3 00
Chester's Catalogue of Minerals. . . . . .8vo, paper, 1 00
                                           Cloth, 1 25
* Crane's Gold and Silver. 8vo, 5 00
Dana's First Appendix to Dana's New "System of Mineralogy". .Large 8vo, 1 00
Dana's Second Appendix to Dana's New "System of Mineralogy."
                                           Large 8vo, 1 50
    Manual of Mineralogy and Petrography. .12mo, 2 00
    Minerals and How to Study Them. 12mo, 1 50
    System of Mineralogy. . . . . .Large 8vo, half leather, 12 50
    Text-book of Mineralogy 8vo, 4 00
Douglas's Untechnical Addresses on Technical Subjects. 12mo, 1 00
Eakle's Mineral Tables 8vo, 1 25
Eckel's Stone and Clay Products Used in Engineering. (In Preparation.)
Goesel's Minerals and Metals: A Reference Book. . . . . .16mo, mor. 3 00
Groth's The Optical Properties of Crystals. (Jackson) (in Press.)
Groth's Introduction to Chemical Crystallography (Marshall). 12mo, 1 25
* Hayes's Handbook for Field Geologists. . . . . .16mo, mor. 1 50
Iddings's Igneous Rocks. 8vo, 5 00
    Rock Minerals. 8vo, 5 00
Johannsen's Determination of Rock-forming Minerals in Thin Sections. 8vo,
                                     With Thumb Index 5 00
* Martin's Laboratory Guide to Qualitative Analysis with the Blow-
    pipe. 12mo, 60
Merrill's Non-metallic Minerals: Their Occurrence and Uses. 8vo, 4 00
    Stones for Building and Decoration. . . . . : 8vo, 5 00
* Penfield's Notes on Determinative Mineralogy and Record of Mineral Tests.
                                      8vo, paper, 50
    Tables of Minerals, Including the Use of Minerals and Statistics of
        Domestic Production. 8vo, 1 0
* Pirsson's Rocks and Rock Minerals. 12mo, 2 0
* Richards's Synopsis of Mineral Characters. . . . . .12mo, mor. 1 05
* Ries's Clays: Their Occurrence, Properties and Uses. 8vo, 5 00
* Ries and Leighton's History of the Clay-working Industry of the United
    States. 8vo, 50
* Tillman's Text-book of Important Minerals and Rocks. 8vo, 00
Washington's Manual of the Chemical Analysis of Rocks. 8vo, 2 00

MINING.

* Beard's Mine Gases and Explosions. . . . . .Large 12mo, 3 00
* Crane's Gold and Silver. 8vo, 5 00
    * Index of Mining Engineering Literature. 8vo, 4 00
         /   * 8vo, mor. 5 00
    * Ore Mining Methods. 8vo, 3 00
Douglas's Untechnical Addresses on Technical Subjects. 12mo, 1 00
Eissler's Modern High Explosives. 8vo, 4 00
Goesel's Minerals and Metals: A Reference Book. . . . . .16mo, mor. 3 00
Ihlseng's Manual of Mining. 8vo, 5 00
* Iles's Lead Smelting. 12mo, 2 00
Peele's Compressed Air Plant for Mines. 8vo, 3 00
Riemer's Shaft Sinking Under Difficult Conditions. (Corning and Peele.)8vo, 3 00
* Weaver's Military Explosives. 8vo, 3 00
Wilson's Hydraulic and Placer Mining. 2d edition, rewritten. 12mo, 2 50
    Treatise on Practical and Theoretical Mine Ventilation 12mo, 1 25

SANITARY SCIENCE.

Association of State and National Food and Dairy Departments, Hartford
    Meeting, 1906. .... 8vo, $3 00
    Jamestown Meeting, 1907. .... 8vo, 3 00
* Bashore's Outlines of Practical Sanitation. 12mo, 1 25
    Sanitation of a Country House. 12mo, 1 00
    Sanitation of Recreation Camps and Parks. 12mo, 1 00
Chapin's The Sources and Modes of Infection. (In Press)
Folwell's Sewerage. (Designing, Construction, and Maintenance.). 8vo, 3 00
    Water-supply Engineering. 8vo, 4 00
Fowler's Sewage Works Analyses. 12mo, 2 00
Fuertes's Water-filtration Works. 12mo, 2 50
    Water and Public Health. 12mo, 1 50
Gerhard's Guide to Sanitary Inspections. 12mo, 1 50
    * Modern Baths and Bath Houses. 8vo, 3 00
    Sanitation of Public Buildings. 12mo, 1 50
    * The Water Supply, Sewerage, and Plumbing of Modern City Buildings.
        8vo, 4 00
Hazen's Clean Water and How to Get It. .... Large 12mo, 1 50
    Filtration of Public Water-supplies. 8vo, 3 00
Kinnicut, Winslow and Pratt's Sewage Disposal. (In Press.)
Leach's Inspection and Analysis of Food with Special Reference to State
    Control .... 8vo,— 7 50
Mason's Examination of Water. (Chemical and Bacteriological). 12mo, 1 25
    Water-supply. (Considered principally from a Sanitary Standpoint).
        8vo, 4 00
Mast's Light and the Behavior of Organisms. (In Press.)
* Merriman's Elements of Sanitary Engineering. 8vo, 2 00
Ogden's Sewer Construction 8vo, 3 00
    Sewer Design. 12mo, 2 00
Parsons's Disposal of Municipal Refuse. 8vo, 2 00
Prescott and Winslow's Elements of Water Bacteriology, with Special Refer-
    ence to Sanitary Water Analysis. 12mo, 1 50
* Price's Handbook on Sanitation. 12mo, 1 50
Richards's Cost of Cleanness. 12mo, 1 00
    Cost of Food. A Study in Dietaries. 12mo, 1 00
    Cost of Living as Modified by Sanitary Science. 12mo, 1 00
    Cost of Shelter. 12mo, 1 00
* Richards and Williams's Dietary Computer. 8vo, 1 50
Richards and Woodman's Air, Water, and Food from a Sanitary Stand-.
    point. 8vo, 2 00
* Richey's Plumbers', Steam-fitters', and Tinners' Edition (Building
    Mechanics' Ready Reference Series). .... 16mo, mor. 1 50
Rideal's Disinfection and the Preservation of Food. 8vo, 4 00
    Sewage and Bacterial Purification of Sewage. 8vo, 4 00
Soper's Air and Ventilation of Subways. 12mo, 2 50
Turneaure and Russell's Public Water-supplies. 8vo, 5 00
Venable's Garbage Crematories in America. 8vo, 2 00
    Method and Devices for Bacterial Treatment of Sewage. 8vo, 3 00
Ward and Whipple's Freshwater Biology. (In Press.)
Whipple's Microscopy of Drinking-water. 8vo, 3 50
    * Typhoid Fever. .... Large 12mo, 3 00
    Value of Pure Water. .... Large 12mo, 1 00
Winslow's Systematic Relationship of the Coccaceæ. .... Large 12mo, 2 50

MISCELLANEOUS.

Emmons's Geological Guide-book of the Rocky Mountain Excursion of the
    International Congress of Geologists. .... Large 8vo 1 50
Ferrel's Popular Treatise on the Winds. 8vo, 4 00
Fitzgerald's Boston Machinist. 18mo, 1 00
Gannett's Statistical Abstract of the World. 24mo, 75
Haines's American Railway Management. 12mo, 2 50
Hanausek's The Microscopy of Technical Products. (Winton) 8vo, 5 00

Jacobs's Betterment Briefs. A Collection of Published Papers on Or-
ganized Industrial Efficiency. 8vo, $3 50
Metcalfe's Cost of Manufactures, and the Administration of Workshops..8vo, 5 00
Putnam's Nautical Charts. 8vo, 2 00
Ricketts's History of Rensselaer Polytechnic Institute 1824–1894.
Large 12mo, 3 00
Rotherham's Emphasised New Testament. ....Large 8vo, 2 00
Rust's Ex-Meridian Altitude, Azimuth and Star-finding Tables........8vo, 5 00
Standage's Decoration of Wood, Glass, Metal, etc.................12mo, 2 00
Thome's Structural and Physiological Botany. (Bennett)..........16mo, 2 25
Westermaier's Compendium of General Botany. (Schneider).........8vo, 2 00
Winslow's Elements of Applied Microscopy.......................12mo, 1 50

HEBREW AND CHALDEE TEXT-BOOOKS.

Gesenius's Hebrew and Chaldee Lexicon to the Old Testament Scriptures.
(Tregelles.) ······························Small 4to, half mor, 5 00
Green's Elementary Hebrew Grammar............................12mo, 1 25

.

Lightning Source UK Ltd.
Milton Keynes UK
UKHW020012160219
337399UK00010B/833/P